FOOD PROBLEMS IN INDIA

FOOD PROBLEMS IN INDIA

RATNA K. BANDILA

ASHISH PUBLISHING HOUSE
8/81, PUNJABI BAGH, NEW DELHI - 110026

Published by :

S.B. Nangia
for **Ashish Publishing House**
8/81, Punjabi Bagh
New Delhi-110026
Tele. 50 05 81
541 09 24

ISBN 81-7024-465-X

1992

Typesetting by :
Computer Codes
Virat Bhawan, Comml. Complex,
Mukherjee Nagar, Delhi - 110 009
Phone : 7241200

Printed at :
Efficient Offset Printers
215, Shahzada Bagh Indl. Complex,
Phase-II, Phone : 533736, 533762
Delhi - 110035

Dedicated to the Loving Memory

of my beloved Father

Late Sri. B. Meshach

FOREWORD

Food supply in an important variable in most of the econometric models which are relevant to the growth problems of developing countries like India. Food problem is a serious problem for the Indian Economy. There has been more or less a continuous imbalance in the economy between aggregate supply and aggregate demand for various types of foodgrains, though the magnitude of the imbalance between demand for and supply of foodgrains has been varying from time to time. The analysis of targets and actual production of foodgrains in Seven Five-Year Plans of India indicates that there are wide fluctuations and divergencies between the targets and achievements of production of foodgrains in the successive Five-Year Plans of India, except in the First Five-Year Plan. The deficiency in the requirement of foodgrains is being met through imports of foodgrains.

The performance of the agricultural sector has received much greater attention than the management of the food problem in the Indian economy. A notable feature of the study of Dr. B. Ratna Kumari is in bringing together, for the first time, the food problem, the growth of the foodgrain economy and the imports of foodgrains of India in a systematic manner. She has brought out the nature of food problem in India, examined the growth of foodgrain economy in India and focused on the role of foodgrain imports of India. Her study points out that imports of foodgrains in India are mostly undertaken to meet the shortfalls in the production of foodgrains in relation to the growing needs of the economy. In recent times, foodgrain imports are also being undertaken to buildup buffer stocks.

By the turn of the century, foodgrain requirement of the country would be around 290 million tonnes. Hence, unless the annual production of the foodgrains in the country is increased to 300 million tonnes, India may be faced with a 'food crisis'. In this context, the author rightly emphasises on the need for the efficient use of the available irrigation facilities in order to improve the availability of foodgrains in the long run. The study of Dr. B. Ratna Kumari is an excellent addition to the literature on Foodgrain Economy of India.

(DR. P. LEELA)
M.A.,Ph.D.,D.A.S.,
Head of the Department of
Corporation and Applied Economics,
Andhra University, Waltair.

(DR. C. RATNAM)
M.A.,(Maths), M.A.(Eco.),
Ph.D.,
Reader and Asst. Principal,
Department of Economics,
Andhra University, Waltair.

Visakhaptanam

ACKNOWLEDGEMENTS

The present study is the manuscript of my Ph.D. Thesis submitted to the Andhra University in 1990. For this my greatest debt is to my Research Director, Dr. P. Leela, M.A., Ph.D., D.A.S., Head of the Department of Co-operation and Applied Economics, A.U., Waltair. I am much benefited by her invaluable guidance, supervision and discussions which helped me to get insights into the problem of my study.

I express my profound sense of gratitude to my Joint Research Director, Dr. C. Ratnam, M.A., (Maths), M.A. (Eco), Ph.D., Reader, Department of Economics and Assistant Principal, Arts and Commerce College, Andhra University, for providing me the benefit of his keen insight into the statistical techniques and immense research.

I am thankful to Prof. G. Parthasarathy, Prof. S.V. Rao, Prof. B. Prasada Rao, Prof. P.V. Sarma and Dr. R.M. Mohan Rao, for their keen interest shown in my research work. I am thankful to all of my teachers in the Department for their kind help extended to me at various stages of my dissertation work.

I express my sincere gratitude to the staff members of the Dr. V.S. Krishna Memorial Library for providing me with the necessary library facilities. I am also deeply indebted to Dr. Seetaramnath and Sri A. Satyanarayana and other staff of the Computer Centre, Andhra University, for their timely and patient help in the processing of the data.

I express my heartfelt thanks to the authorities of LWF, in particular Dr. G. Raja Ratnam, Executive Secretary, UELCI, Madras and Ms. Verena Mosciatti, Administrative Asst., LWF, Geneva for awarding the LWF Fellowship to me and enabling me to undertake the research work. I am thankful to Dr. Rev. K. Nathaniel, President, AELC for sponsoring my candidature for the LWF Fellowship.

I am also thankful to Prof. M. Jagadeeswara Rao, Coordinator, School of Economics, the office staff of School of Economics and the office staff of Department of Applied Economics for their generous help and the facilities provided to me.

I am much grateful to M/s. Ashish Publishing House, New Delhi for publishing this book in a very special manner.

I would like to place on record the affectionate encouragement received from my parents, Smt. B.M. Angeleena and Late Sri B. Meshech and my brothers and sisters. Lastly, no words can adequately express my deep sense of gratitude to my beloved husband, Mr. G. Paul Prabhakar, Forest Range Officer, who has given me inspiration, encouragement and continuous help in abundance throughout my research work. I extend my heartfelt thanks to my children, G. Dheeraz, G. Deepthi and G.M. Sundeep for sharing more than their fair amount of family burdens to facilitate the availability of as much time as I needed, to complete my research work. My thanks are also due to my parents-in-law Mrs. G. Sowbhagyamma and Rev. G. Raja Ratnam and my brothers-in-law and sisters-in-law for their moral support.

RATNA K.BANDILA

POST-DOCTORAL FELLOW
DEPT. OF COOP. & APP. ECONOMICS
ANDHRA UNIVERSITY, WALTAIR

Visakhapatnam

CONTENTS

LIST OF TABLES

CHAPTER 5

APPENDIX TABLE

LIST OF GRAPHS

LIST OF DIAGRAMS

CHAPTER 1

INTRODUCTION

1.1. Agriculture in the Indian Economy

COUNTRIES with relatively large atricultural sectors depend more on agricultural growth to susatain their development than do the others.[1] As pointed out by Johnson and Mellor,[2] accelerated agricultural development stimulates economic growth in a number of ways. Firstly, economic development is identified by a substantial increase in the demand for agricultural products. At a minium, farm production must be increased rapidly enough to keep pace with population growth. Secondly, agriculture must generate export surpluses in order to earn the foreign exchange with which to finance the imports. Thridly, agriculture is also a supplier of production factors such as capital and labour. Fourthly, agriculture also provides many of the raw materials for industry. Kuznets[3] refers to the 'product', 'market' and 'factor' contributions of agriculture to econonic development. Further, agriculture in one from or other is called upon to finance a significant part of the investment for the expansion of industry, transport and other sectors as well.

Thus, agriculture contributes substantially to the economic development of the developing countries. Rostow[4] observes that agriculture plays a distinct, but multiple, role in the transitional

---

1. M. Yudelman, *Agricultural Development and Economic Integration in Latin America*, George Allen and Unwin Ltd., London, 1970, p.67.
2. B.F. Johnson and J.W. Mellor, "The Role of Agriculture in Economic Development", *American Economic Review*, Vol. L1, No.4, September 1961, pp. 556-581.
3. S. Kuznets, "Economic Growth and the Contribution of Agriculture: Notes on Measurement", *International Journal of Agrarian Affairs*, Vol. 3, No.2, April 1961, pp. 56-75.
4. W.W. Rostow, *Stages of Economic Growth*, Cambridge University Press, London, 1962.

process of the 'take-off' into self-sustained growth. Schultz[5] also opines that agriculture has become one of the sectors that can contribute much to economic growth.

High priority has been given to agriculture in the First Five-Year Plan of India, in view of the important role played by agriculture in promoting and sustaining a rapid rate of economic growth in the country. Indian economy is still predominantly agricultural in nature. The contribution of the agricultural sector to the economic development of the country is indicated by its contribution to national income, employment, industrial development and foreign exchange earnings of the country. Agriculture and allied activities contributed about 33 per cent to the Gross Domestic Product of India in 1987-88. Agriculture is the mainstay of the rural people in India. Agricultural sector provides livelihood to about 75 per cent of the population of the country. It also provides raw materials like cotton, jute and sugar to a number of industries, besides providing a wide market for the industrial products. Agricultural and allied exports accounted for about 19 per cent of the total exports of India in 1987-88.[6]

Thus, agriculture is an important segment of the Indian economy. With the help of modern science and technology, it will continue to play a prominent role in lifting the country out of the morass of poverty and unemployment. It has become almost axiomaqtic that drawbacks in the agrarian sector act as a serious obstacle to economic development in general.[7]

An increase in agricultural output is important for providing the necessary food for the growing population, raw materials for the industry and markets for industrial products. About one-quarter of economic activity in India is involved in the production of foodgrains.[8] Foodgrain crop plays a crucial role in the agricultural economy of India. Foodgrains produced in India can be broadly

---

5. T.W. Schultz, "A New Era for Agriculture in Economic Growth", *Indian Journal of Agricultural Economics*, Vol. XIV, No. 4, October-December, 1959, p.37.
6. Based on data from GOI., *Economic Survey 1988-89*, 1989, pp. S-74 and S-75.
7. M.L. Dantwala, "Agrarian Structure and Economic Development", *Indian Journal of Agricultural Economics*, Vol. XV, No.1, January-March 1961, pp.10-25.
8. J. Wall, "Foodgrain Management: Pricing, Procurement, Distribution, Imports and Storage Policy", in World Bank, *India: Occasional Papers*, Staff Working Paper No.279, November 1978, p.43.

classified into three major groups, *viz.*, (*a*) cereals, (*b*) coarse grians/millets and (*c*) pulses. India's main cereal crops are: rice and wheat. Jowar, bajra, barley and maize constitute the important coarse grains of the country. Pulses consist of : redgram, blackgram, bengalgram, tur, masur, arhar, mung and other minor crops.

1.2. Importance of Food in Indian Diet

Food is one of the basic needs of human existence. Cereals like rice, wheat and millets constitute the basic food of human beings. The dietary surveys conducted by the Indian Council of Medical Research indicate the importance of cereals in the Indian diet for the different social classes like agricultural workers, industrial workers and others.[9] The distribution of family expenditure on food also indicates the importance of cereals in the diet. Cereals and pulses constitute a major part of the food budget of majority of the population in India.[10]

There are considerable variations in the proportion of the components of the diet. In States like Punjab, wheat is the main cereal. In the East and South, rice is the staple cereal. In Central and West India, millets are the main cereals. Consumption of cereals is found to be higher in the rural than in the urban areas. In spite of these variations, cereals play a predominant role in the Indian diet.

Cereals account for about two-thirds of India's total calorie intake and somewhat smaller proportion of the protein content of the average diet. Of the cereals, roughly one-half is rice, one-sixth is wheat and the remaining one-third consists of millets.[11] The importance of pulses in the Indian economy is on account of two factors: (*a*) pulses offer an effective solution to the 'protein problem' and (*b*) they have a useful role to play in replenishing soil fertility as they add more nitrogen to the soil than fertilisers.[12]

---

9. See, S.C. Joseph, *Food Policy and Economic Development in India,* Allied Publishers Pvt. Ltd., Bombay, 1961, p.3.
10. A.K. Mitra, *Food Problem in India: A Quantitative Study in the Regional Perspective,* Anu Books, Meerut, 1986, p.XIV. In 1950-51, the percentage of expenditure on cereals was found to be 85.3, by the All India Agricultural Labour Inquiry. See GOI., *All India Labour Inquiry 1950-51*, p.41.
11. U.N., F.A.O., Commodity Bulletin Series, *The Economic Relationships Between Grains and Rice*, Report 39, USA, 1965, p.7.
12. P.C. Bansil, *Agricultural Problems of India*, Oxford and IBH Publishing Co., New Delhi, 1981, pp. 374-75.

1.3. Food and Economic Development

Food supply plays an important role in economic development. This fact is emphasised in most of the models of economic growth. The classical economists have assigned a central place to food in their growth models, which are relatively more relevant to the growth problems of the developing countries.

According to Adam Smith,[13] it is the amount of food, produced in the country, which limits the size of labour force or population in a country. According to him, food supply can be increased through improvements in agriculture. However, Ricardo[14] maintained that food supply becomes a limiting factor in economic progress because of its influence on wages. Ricardo's model has considerable relevance to the food problem of the developing countries. These countries do not have the necessary foreign exchange to import food from abroad. Hence, they have to necessarily increase domestic production of food in order to see that their developmental efforts are not thwarted by rising prices of foodgrains and inflation.

Food supply constitutes an important source of capital for the underdeveloped countries. The most important source of capital formation in the underdeveloped countries is the surplus labour in agriculture which could be used for the production of wealth in the country. Such a transfer of labour from agriculture to non-agricultural occupations is influenced to a large extent by the availability of food surplus from the agricultural sector to feed the non-agricultural population. Hence, increasing the market supplies of foodgrains becomes important for starting the process of growth in underdeveloped countries. Food surplus is, thus, of strategic importance to the economic development of the underdeveloped countries. From this it follows that highest priority should be given to the increase in food production in the plans for economic development of the underdeveloped countries.

Food sector plays a crucial role in the development of the Indian economy. If the food sector is lagging, industrialisation programmes would not gather momentum. The technological developments in

---

13. Adam Smith, *Wealth of Nations*, Modern Library Edition, New York, 1913, p. 163.
14. David Ricardo, *Principles of Political Economy*, Sraffa Edition, London, p.99.

agriculture tend to suggest that the food sector is likely to develop at a rapid rate and reconcile the objective of growth with stability and equity. One aspect of the food problem is the problem of hunger, which is an integral component of poverty. Hence it would not be proper to judge the state of food problem entirely on the basis of market indicators such as prices and market arrivals. The problem of hunger would call for more direct measures to treat certain minimum needs of food of the poor as the first priority on the food that is produced.[15]

1.4. Nature of the Food Problem in India

Food problem is a serious problem for the Indian economy. Excluding a few years of exceptionally good harvest, Indian economy is faced with the problem of deficiency in the requirement of foodgrains. There has been more or less a continuous imbalance in the economy between aggregate supply and aggregate demand for various types of foodgrains, though the magnitude of imbalance between demand for and supply of foodgrains varies from time to time.

As a nation, India is short of food both in absolute terms as well as in comparative terms. Shortfall in calories is treated as under-nutrition. Mal-nutrition means a nutritionally unsound composition of the foods from which the calories are derived. About 250 million people of India are either under-nourished or mal-nourished or both.[16] This gives in capsule form, the nature of the food problem in India, apart from the intensified problems arising out of shortfalls in production. The long-term food problem involves questions of production, availability of domestic supplies distribution over time between States, effective demand and its relation to income and employment and of the needs of vulnerable groups of the society.

Sukhatme made a special study of the Indian food problem from the nutritional angle. Though his findings relate to the years 1958-60, they deserve attention. According to Sukhatme, at least one in every four and probably one in every three of India's population is under-

15. V.M. Rao, *Second India Studies: Food*, The MacMillan Company of India Ltd., Delhi, 1975, pp.2-3
16. V.K.R.V. Rao, *Food, Nutrition and Poverty in India*, Vikas Pub. House Pvt. Ltd., 1982, p. 20.

nourished, which means that they do not have a minimum quantity of food in terms of minimum calories.[17]

The per capita per day availability of food in India in comparison with the underdeveloped as well as the developed countries is shown in Table 1.1. It is evident from the Table that the overall Index of total food availability per capita per day in India is 54 compared to 70 for the underdeveloped countries, 179 for the developed countries and 100 for the World as a whole. In other words, India's total Food Index is about half that of the World Food Index and less than one-third that of the developed world.

The position is not very encouraging in terms of total proteins, with 51 grammes for India as compared to 60 grammes for the underdeveloped World, 90 grammes for the developed World and 68 grammes for the World as a whole. The availability of food per capita per day is 1,970 calories for India compared to 2,190 calories for the underdeveloped countries, 3,060 calories for the developed countries and 2,420 calories for the World as a whole.

Food problem has two aspects namely production and distribution which are closely interconnected. The stated objectives of food policy are : (*a*) to achieve self-reliance in production, (*b*) to ensure equitable distribution and (*c*) to bring about price stability.[18] The problem of food is not one of merely making up the deficiency in supplies as ensuring their better distribution by States or making special arrangements for distribution to the vulnerable classes. There is also the question of prices, incomes and income elasticity of demand for food.[19]

The nature of food problem in India has to be assessed in terms much deeper than temporary and marginal deficits in foodgrains supply. The rapid rise in population together with growing imports of

17. P.V. Sukhatme, *Feeding India's Growing Millions*, Asia Publishing House, 1965, quoted from V.K.R.V. Rao, *Food, Nutrition and Poverty in India, op. cit.*, p. 20.
18. See GOI., *The Foodgrains Policy Committee*, 1966.
19. V.K.R.V. Rao, *op. cit.*, pp. 24-26.

Table 1.1 : Per Capita Availability of Food in India and other countries (in grammes per person per day at retail level*)

Items	India	Under Developed countries	Developed countries	World
1	2	3	4	5
Cereals	375	393	328	370
Starchy Roots	30	229	316	227
Sugar	45	26	88	47
Pluses and Nuts	65	50	16	42
Fruits and Vegetables	80	191	362	227
Meat	4	37	152	67
Fish	7	28	34	27
Eggs	1	5	33	12
Milk and Milk Products	140	64	573	228
Fats and Oils	11	12	47	22
Calories	1,970	2,190	3,060	2,420
Animal Protein	6	10	44	20
Total Protein	51	60	90	68
Fats	27	36	106	56
Overall Index of Animal Food	30	47	233	100
Overall Index of Total Food	54	70	179	100

Note : * Retail level means the foodgrains and other foods available for consumption and excludes that part of production which is not available for consumption due to deed, feed and wastage.

Col. 3 includes Far East, Near East, Africa and Latin America excluding River-Plate countries.

Col. 4 includes Europe, North America, Oceania and the River-Plate countries.

Source : V.K.R.V. Rao, *Food, Nutrition and Poverty in India*, Vikas Pub. House Pvt. Ltd., New Delhi, 1982, p. 21.

foodgrains focused attention on the food problem,. The whole food problem came to be viewed in the Malthusian setting as a race between population and food production.[20]

The food problem in India is essentially a foodgrains problem.[21] This is primarily because of the importance of foodgrains in the

20. R.P., Sinha, *Food in India : An Analysis of the Prospects for Self-sufficiency by 1975-76*, Oxford University Press, Bombay, 1961, p. 2.
21. S.S. Madalgi, *Population and Food Supply in India*, Lavani Publishing House, Bombay, 1970, p. 4.

Indian diet. Shortage of foodgrains might lead to starvation for a large part of the low-income groups of people in the economy. Uncertainity in respect of the quantitative adequacy might lead to a meal or no meal for certain sections of the population. Hence, the study of foodgrains becomes important.

1.5. Brief History of India's Food Problem

The heart of the World food problem lies in East Asia, as about one-half of the World's population lives in East Asia. Of this about one-fourth is underfed.[22] The danger of loosing the race between increase in the demand for food and development of agriculture is most prominently found in this part of the World, including India.[23] The leading elements in this situation are : (*a*) political revolution, which brought about freedom to a large number of colonies and (*b*) demographic revolution resulting in population explosion. The high rates of population growth together with increase in the demand for food due to the increase in incomes makes the food situation acute in countries like India.

Food problem in India has become quite complex since the Second World War. The fall of Burma in 1942 and the Bengal Famine in 1943 highlighted the food scaricity conditions in the Indian Union. The Bengal Famine may be regarded as a landmark in the history of food and famine problems in India. It focused attention on the precarious food situation in the country. In spite of the frequent famines faced by the country in the 19th century, India's food policy was not developed till 1943, when there was a gradual realisation on the part of the Government that measures to increase productivity are as important as measures to prevent calamities. As a result, the 'Grow More Food Campaign', was launched, as a part of the food policy of the country in 1943.

The Indian economy experienced crop failures in 1945 creating near famine conditions. The partition of the country in 1947 further weakened the supply of food to the Indian economy. "India received 82 per cent of the population with 75 per cent of the production of the principal foodgrains and 69 per cent of the net irrigated area."[24] In

---

22. U.N., F.A.O., *6 Billions to Feed*, Rome, 1962, p. 3.
23. B.M. Bhatia, *India's Food Problem and Policy Since Independence*, Somaiya Publications Pvt. Ltd., Bombay, 1970, pp. 26-30.
24. S.C. Joseph, *op. cit.*, p. 2.

1949, the 'Grow More Food Campaign' received a fresh impetus with the emphasis placed on self-sufficiency in food. "Though Government—sponsored efforts for increasing the production of foodgrains started under the 'Grow More Food Campaign' in 1943, to meet the shortage of food caused by the stoppage of imports of rice from Burma, self-sufficiency as an objective of food production policy was announced for the first time towards the end of 1948.[25] The campaign also came to be incorporated in the First Five-Year Plan (1951-56), which gave high priority for agriculture and irrigation.

With the emphasis on active measures to raise the standard of living of the people, India's food problem came to be viewed in a new perspective. In a growing economy, the food consumption depends not merely on population growth alone, but also on the rate of growth of per capita income and the changes and the technological improvements in the utilisation of natural resources.[26] Food problem in India is not considered merely the result of population explosion alone but also due to economic development.[27] In developing countries, the demand for and supply of foodgrains are very often in disequilibrium after the initiation of the development process.

Agricultural development received a low priority in the Second and Third Five-Year Plans. Consequently, agricultural development started slowing down and reached a stagnation point in 1960-61. There were two successive crop failures in 1965-66 and 1966-67. This focused on the need to increase domestic production of food. The New Agricultural Strategy, popularly known as Green Revolution was adopted in 1965-66, consisting of the use of HYV seeds, fertilisers, pesticides and irrigation.

The rapid rise in population and purchasing power in the hands of the people led to increase in the demand for food, which could not be matched by the production of foodgrains, resulting in rise in prices of foodgrains and frequent occurances of periods of scarcity. The economy also experienced shortages in respect of different types of

---

25. J.S. Sharma, "India—A Drive Towards Self-sufficiency in Foodgrains", *American Journal of Agricultural Economics,* Vol. 60, No. 5, December 1978, p. 859.
26. S.S. Madalgi, *op. cit.*, p. 3.
27. B.M. Bhatia, *op. cit.*, p. 23.

foodgrains as well as shortages in different regions. Governments, at the Centre as well as at the State level, have undertaken various measures including structural, technological and economic changes to bring about adjustments between the supply and demand for foodgrains. These measures have resulted in the easing of the problems on the food front to some extent in the eighties.[28]

1.6. Foreign Trade and the Role of Imports

"Trade is essentially an international transformation of commodities, inputs and technology which promotes welfare in two ways. It extends the market of a country's output beyond national frontiers and may ensure better prices through exports. Through imports, it makes available commodities, inputs and technology which are either not available or are available only at higher prices, thus, taking consumers to a higher level of satisfaction."[29] The foremost principle of foreign trade, *viz.*, 'the law of comparative costs', signifies that what a country exports and imports is determined not by its character in it solution but only in relation to those of its trading partners.[30] According to Samuelson "foreign trade offers a consumption possibility frontier that can give us more of all goods than can our own domestic production-possibility frontier."[31] The extension of foreign trade, according to Ricardo "will very powerfully contribute to increase the mass of commodities, and therefore, the sum of enjoyments."[32] This will we true for each trading nation. In modern terminology, "trade is a positive sum game."[33]

International trade is an activity of strategic importance in the development process of a developing economy.[34] Underdeveloped countries are concerned with their international trade position,

---

28. A.K. Mitra, *op. cit.*, p. 13.
29. J. Viner, *International Trade and Economic Development*, Clarendon Press, Oxford, 1952, p. 146.
30. C.P. Kindleberger, *Foreign Trade and the National Economy*, Yale University Press, New Haven, 1962, pp. 3-4.
31. P.A. Samuelson, *Economics*, McGraw Hill, Kogakusha Ltd., 1973, p. 643.
32. David Ricardo, *Principles of Political Economy and Taxation*, The MacMillan Press Ltd., London, 1817, p. 107.
33. G.M. Merier, *International Economics : The Theory and Policy*, Oxford University Press, New York, 1980, p. 57.
34. V.R. Panchamukhi, *Trade Policies of India : A Quantitative Analysis*, Concept Pub. Co., Delhi, 1978, p. 18.

because for all of them, international trade is vitally important as a source of supply of technological know-how, skills, capital, machinery and implements, which are essential for their economic development.[35]

Besides these imports, a developing economy is also required to import consumer goods, which are in short supply at home during the period of industrialisation. Such imports are anti-inflationary in nature because they reduce the scarcity of consumer goods. It is, therefore, inevitable that during the early years of development, imports may increase at a fast rate. Hirshman indicated how the growth of imports precedes the establishment of a domestic industry by demonstration the existence of a market.[36] Manmohan Singh[37] states that import needs generally tend to increase rapidly in the early stages of economic development, on account of the structural imbalances and rigidities, which characterise many developing countries.

Imports into developing countries as well as the geographical and commodity composition of these imports, are greatly influenced by the rate of economic growth and structural changes in these economies.[38] Kindleberger points out that the process of economic transformation is usually followed by rise in the proportion of imports to national income due to increase in the needs of the people and improvements in consumption and industrial requirements of the economy. Thus, economic growth has an effect on imports in their turn influence economic growth, which involves a process of raising the productive capacity of the economy.[39]

1.7. Imports and Economic Development in India

Planning in India is intended to promote a rapid rise in the standard of living of the people by efficient allocation of the resources

35. G. Haberler, "Terms of Trade and Economic Development" in H. Ellis (ed.), *Economic Development of Latin America*, St. Martin's Press, New York, 1961, pp. 275-97.
36. A.O. Hirschman, *The Strategy of Economic Development*, Yale University Press, New Haven, 1958, pp. 120-122.
37. M. Manmohan Singh, "Trade Polices and Economic Development", in *Economic Development in South Asia* (ed.), by E.A.G. Robinson and M. Kidron, proceedings of a conference held by the International Economic Association at Kandy, Ceylon, MacMillan, St. Martin's Press, London, 1970, pp. 455-59.
38. B. Balassa, *Trade Prospects for Developing Countries*, Yale University Press, 1964, pp. 67-99.
39. C.P. Kindleberger, *Foreign Trade and the National Economy, op. cit.*, pp. 9-11.

of the country. In the massive efforts, imports as well as exports play an important role. Imports have been increasing both in absolute terms as well as relative to national income, particularly since 1956. In India, the quantitative significance of foreign trade is relatively low. The share of total imports in the Gross Domestic Product of India in 1987-88 is found to be 7.64 per cent. However, to appreciate the qualitative significance of imports, it is necessary to underline the crucial role that imports play in the country's industrial and economic growth. In the Indian context, the industrial output has become so much linked to the foreign-supplied inputs that a failure to meet these needs soon results in under-utilisation of industrial capacity and the kind of industrial capacity and the kind of industrial stagnation hypothesised by the critics of import substitution.[40]

Persistent balance of payments difficulties have constrained and sometimes even frustrated India's efforts to generate self-sustained economic growth. 'Efficient import-management' has become the necessary condition to bring about adjustment in the balance of payments not only in the short-run but also in the long-run. In order to estimate import requirements, planners require specification of import structure and import substitution process. Irrespective of the policies that are adopted, imports have to exhibit some kind of macro-relationship with domestic economic variables such as industrial production, investment, inventory holdings, relative prices, foreign exchange receipts, etc.[41]

Expansion in imports may take place in three ways : (*a*) increase in the imports of materials not available in India ; (*b*) increase in the imports of materials, equipment not made in India ; and (*c*) increase in the imports of those items produced in India, but where demand exceeds domestic output. The first would depend on the growth of demand consequent upon the growth of industries using them, whereas in the third case, there is a spill-over of domestic demand into imports. Consistent with the availability from within the country, it would, therefore, be necessary to evolve a pragmatic import strategy which would meet the essential import requirements of agriculture, industry, infrastructure and other sectors of the economy. It is necessary that a supportive import policy is pursued over a period of time.

---

40. S.K. Patel, *Quarterly Econometric Analysis of Imports in India*, Hindustan Publishing Corporation, Delhi, 1981, pp. 6-23.
41. *Ibid.*, p. 1.

1.8. Importance of Foodgrain imports in India

Before the beginning of the First World War, India enjoyed a small foodgrain export surplus,[42] with exports of foodgrains exceeding imports. In the year 1921, India became a net importer of foodgrains. This may be seen from Table 1.2. It is evident from the Table that net imports of foodgrains of India on an average, increased from 1.6 lakh tons during the period 1920-21 to 1924-25 to 13.8 lakh tons during the period 1935-36 to 1939-40. However, the net imports of foodgrains were relatively small and accounted for an insignificant portion of the country's total food supply until the beginning of the Second World War.[43]

Table 1.2: Imports and Exports of Foodgrains of Undivided India

Five Year Period	Quantity (in lakh tons)		
	Imports	Exports	Net Imports
1915-16 to 1919-20	11.0	15.9	–4.9
1920-21 to 1924-25	11.4	9.8	1.6
1925-26 to 1929-30	15.9	8.3	7.6
1930-31 to 1934-35	18.4	5.7	12.7
1931-36 to 1939-40	20.7	6.9	13.8

Source : B.M. Bhatia, *India's Food Problem and Policy Since Independence*, Somaiya Publications Pvt. Ltd., Bombay, 1970, p. 129.

The Bengal Famine led to the adoption of a system of controls which, for its successful operation, necessitated resort to imports of foodgrains every year. The procurement of internal supplies failed to keep pace with the growing quantity of public distribution of grains and the gap between the two was bridged by imports. Imports of foodgrains thus came to be built into the whole system of food

42. India's food problem before World War I was mainly the dependence of the country on Burma for the supply of over 2 million tons of rice annually to supplement the country's domestic production. See B.M. Bhatia, *op. cit.*, p. 130.

43. Imports of rice constituted 5 per cent of India's requirements. Rice was mainly imported from Burma. Towards the end of the Second World War, rice shortage became quite acute and visible in terms of rising prices. See S.C. Joseph, *op. cit.*, p. 21.

management in India and became hence forth a permanent feature of India's food economy.[44]

Starting from 1943-44, imports of foodgrains were obtained on Government account only. The magnitude of internal procurement and the internal food supplies have an influence on the volume of net imports. The net imports of Cereals into the Indian Union are shown in Table 1.3. It is evident from the Table that, the net imports of Cereals of India increased from 1.6 million tonnes in 1944 to a high level of 4.7 million tonnes in 1951, but declined to 3.9 million tonnes in 1952 and 2.0 million tonnes in 1953. Imports of foodgrains have thus come to stay involving a heavy annual drain on the limited foreign exchange resources of the country.[45] With the growth in population and increase in incomes consequent to economic development, the demand for food, particularly foodgrains has been growing rapidly and outrunning the supply, necessitating imports to meet the growing requirements.

Imports of food, fertiliser and petroleum products together account for a proportion of the total import bill of India. The share of these bulk imports in total imports increased from 21.99 per cent in 1961-62 to 37.0 per cent in 1985-86. Imports of Food and Live Animals chiefly for Food accounted for 4.34 per cent of the total Imports of the country in 1985-86. Cereals and cereal preparations constitute an important part of imports of Food and Live Animals chiefly for Food. They accounted for 12.9 per cent of total Food imports of India in 1985-86.

Foreign trade in foodgrains is a Government of India monopoly. The Government of India acting through the Food Corporation of India (FCI) decides on how much to import of what types of grain. Since 1971, only a small quantity has been made available on concessional terms and most grain imports have been made available

---

44. The Foodgrains Policy Committee of 1947 concluded that the real solution is not import or controls on procurement and distribution, but only a substantial increase of domestic production within the earliest possible time to solve the Indian food problem. See B.M. Bhatia, *op. cit.*, p. 131.
45. This means that the policy of controls has failed in the achievement of the objective of self-sufficiency. According to the Foodgrains Procurement Committee of 1950, the cause for this failure was that procurement had lagged behind Government commitments for public distribution, which in turn had led to the increasing dependence on imports. See B.M. Bhatia, *op. cit.*, p. 133.

on cash terms or with deferred payment, with little grant element.[46] In recent years, almost all of the grain imported has been wheat. However, India does import small quantities of rice and other grains also.

Table 1.3 : Net Imports of Cereals into the Indian Union

Three year ending (Average)	Rice	Wheat	Total Cereals (including other grains)
1944	0.2	1.3	1.6
1945	0.5	1.3	1.8
1946	0.6	1.4	2.6
1947	0.8	0.9	2.7
1948	0.9	1.3	2.8
1949	0.8	2.2	3.7
1950	0.4	1.4	2.1
1951	0.7	3.0	4.7
1952	0.7	2.5	3.9
1953	0.2	1.7	2.0

Source : *Food Situation, 1939-53*, p. 14, quoted from S.C. Joseph, *Food Policy and Economic Development in India*, Allied Pub. Pvt. Ltd., Bombay, 1961, p. 59.

The quantitative imbalance in the requirement of foodgrains has led to imports of foodgrains in India. However, in spite of the important role played by foodgrain imports in the Indian economy, there has been a relative neglect of the study of foodgrain imports in relation to the food problem of India. Hence, there is a need for undertaking a systematic study of the significance of imports of foodgrains of India in view of the chronic nature of the food problem in India. In this context, a review of some of the earlier studies on food problem and foodgrain imports of India is taken up in the next Chapter.

46. U.N., World Bank, *India : Occasional Papers*, World Bank Staff Working Paper, No. 279, November 1978, p. 58.

CHAPTER 2

REVIEW OF LITERATURE AND APPROACH OF THE STUDY

THOUGH India is World's fourth largest producer of foodgrains, India supplements her foodgrain production with imports, due to the imbalance between food requirements and food output. Imports of foodgrains are included in the bulk imports of India and come under the Standard International Trade Classification (SITC) Section 'O'. Interesting developments have been noticed in the foodgrain imports of India over time. Hence, an analysis of foodgrain imports in relation to the food problem of India would throw considerable light on the contribution of foodgrain imports to the Indian economy.

A Brief review of some of the major studies relating to food problem of India in general and foodgrain imports in particular is presented in Section 2.1. The limitations of the earlier studies are presented in Section 2.2. Section 2.3 is devoted to the specification of the objectives of the present study, while the methodology adopted in the present study is examined in Section 2.4. The nature and sources of data are indicated in Section 2.5. Section 2.6 provides the plan of the study.

2.1. Survey of Earlier Studies

There is a fairly vast literature on various aspects of the food problem of India. Some of the major studies on India's food problem are reviewed in this Section to provide a proper perspective for the analysis of foodgrain imports in relation to the food problem of India.

S.C. Joseph[1] in his study has undertaken a critique of food controls, as administered in India during the period 1939-54. It

1. S.C. Joseph, *Food Policy and Economic Development*, Allied Publishers, Bombay, 1961.

comprises of both a historical account of the various measures of supply and distribution of foodgrains undertaken by the Government during the period under study and an economic interpretation of the course of action which would have better fitted India's economic situation. Food controls were instituted under the stress of a total War and confronted with problem of foodgrain shortages, they were allowed to continue in the post-War years with progressive vigour. The author concludes that food controls were neither effective nor successful in solving the country's food problem. Considering the marginal nature of the foodgrains deficit (about 5 per cent of production) and the general composition of the diet, the author considers free trade in foodgrains as a better solution, modified by State action. According to him, there is need for a qualitative approach to the food problem, *i.e.*, changes in food habits and increased production of quality foodstuffs as a permanent solution of India's food problem on the basis consistent with the need for economic development. However, the study is basically a historical—economic study. Further, though the study examined some trends in imports of rice and wheat, it mainly relates to a very early period *viz.*, 1939-54.

In his study, R.P. Sinha[2] has been mainly concerned with projections of the domestic demand and supply of food in India for the year 1975-76 and with an analysis of the prospects of self-sufficiency by 1975-76. On the demand side, he considers the growth of population and per capita income elasticity of demand for food. Assuming a rectilinear relationship between total expenditure and the expenditure of food, Sinha estimates income elasticity of demand for food, based on cross-section data. The author maintains that self-sufficiency in food can be achieved only if, along with a radical change in food habits, there is a reallocation of agricultural resources, a considerable extension both of the cultivated area and of double-cropping, and an increase in productivity per hectare. In the absence of these conditions being fulfilled, India will have to import a subtantial quantity of food to meet the increasing demand. The need for imports of food, the rise in investments, the increase in interest payments on foreign capital lead to dependence on foreign trade of the economy. Sinha suggests a gradual shift of emphasis within

2. R.P. Sinha, *Food in India*, Oxford University Press, London, 1961.

agriculture from the production of cereals to the production of dairy products and a similar switch over from the import of foodgrains to the import of fodder, which is cheaper. However, the author has not examined the trends in the imports of foodgrains in his study.

Bepin Behari[3] attempts to study the influence of changes in agricultural organisation on the composition, pattern and magnitude of Indian imports and the interconnection between imports and economic development. Food imports are included in the category of consumer goods imports. He makes an attempt to bring together data relating to domestic production of foodgrains and their imports during the period 1947-48 tyo 1962-63. The trends indicate that despite its emphasis on agricultural development, the country has remained basically a 'deficit one'. The rapidly growing population and the dependence of agriculture on monsoons are held to be responsible in preventing India from attaining self-sufficiency in foodgrains. Based on the method of interpolation, the author makes projections for imports for the year 1975-76. Imports of food items are estimated to decline substantially and domestic production of foodstuffs is estimated to almost attain the level of self-sufficiency by the end of the Third Plan. The author himself indicates that these projections could not be realised due to a fall in agricultural production. However, the author has not examined India's food problem in-depth. Further, the study relates to an early period 1947-48 to 1962-63 and is basically analytical in nature.

A developmental approach to import policy was initiated in India during the Second Five-Year Plan (1956-61). Against this framework, the ESRF study[4] examines the changes in the pattern of imports and attempts to define the broad trends in import utilisation. The structural changes in imports are analysed by classifying imports into three categories, *viz.*, (*a*) capital goods, (*b*) producer goods and (*c*) consumption goods, which include food and other consumption imports. Food accounts for roughly one-fifth of total imports, though there are wide fluctuations. The study points out that, though the overall progress in 'self-reliance' seems to be generally satisfactory, agriculture is a striking exception. The analysis of the study reveals

3. B. Behari, *Imports in a Developing Economy*, Vora & Co., publishers, Bombay, 1965.
4. E.S.R.F. *Structure of Indian Imports 1957-1964*, Economic & Scientific Research Foundation, Paper 4, New Delhi, 1966.

that the Indian economy has been depending increasingly on imports for its requirements of food as well as of agricultural inputs such as fertilisers. The fact that food imports alone, among consumption imports, have not declined is attributed to the failure of the development policy. However, the study has not examined the trends in food imports in the context of India's food problem. Further, the study relates only to the period 1957-65.

Imports of India have been undergoing rapid changes since Independence, Diwan[5] studied the import structure and policy of India for the period 1951-1962. He classified imports broadly into three groups, *i.e.*, (*i*) cereals, (*ii*) 'capital and raw materials' and (*iii*) others. He points out that while increasing pressure of population necessitated increase in food imports, exphasis on rapid industrialisation and the use of modern technology in the Second Plan, as against the emphasis on agriculture in the First Plan, created a large demand for imports of the group 'capital and raw materials'. However, the author has not examined India's food problem in relation to cereal imports.

Bhagwati and Desai[6] have undertaken an analysis of India's trade and industrial policies and made a systematic study of trends in industrialisation, economic growth and import substitution. In their study, developments in the import policy of the country have been traced out since 1940 and effects of the import control policies on the economy have been analysed in detail, with a view to suggest a sound import policy for the Indian economy. They noted that rationally planned development is the only means to achieve sustained and efficient growth of the economy. In examining India's economic growth since Independence, they found that the overall performance of Indian agriculture, over a sustained period of fifteen years, is remarkable by international standards, even though it has been inadequate to match the demand arising from growth in population and incomes, thus leading to food imports. However, the authors have not dealt at length either the food problem or foodgrain imports of India, as their emphasis was more on India's experience with the trade and industrialisation policies relating to the period 1951-56.

---

5. R.K. Diwan, "Import Structure and Policy", *Foreign Trade Review*, Vol. III, No. 1, April-June, 1968, pp.1-12.
6. J.N. Bhagwati and P. Desai, *India: Planning for Industrialisation*, Oxford University Press, London, 1970.

As food prices are pace-setters for other prices, in a developing economy like India their stabilisation would go a long way in removing obstacles to a steady rate of economic development without inflation. In this context, Madalgi[7] in his study examines the various aspects of the food problem in India like the demand—supply gap, Government operations in foodgrains, foodgrain consumption levels, self-sufficiency and foodgrains distribution policy. According to Madalgi, without a broad idea of the magnitude of rural and urban demand for foodgrains, policy decisions regarding food management cannot be made properly on the basis of average per capita availability alone. In his study, Madalgi aims at suggesting a technique of analysis of foodgrains supply-demand situation on sectoral basis. He also attempts to measure the actual foodgrains consumption levels of different sections of the population, in order to provide the policy maker with some material, for working out a rational foodgrains distribution arrangement. However, he has not examined the determinants of imports of foodgrains and the food problem in relation to foodgrain imports, in detail. Further, the study relates only to the period 1950-51 to 1967-68.

In his study, Bhatia[8] deals with India's food problem and policy since Independence, against a historical background of the food problem in the pre-Independence period. According to Bhatia, the Indian food policy since the Second World War has been caught up in a vicious circle. Inflation or fear of inflation leads the policy makers to control prices and impose various restrictions on trade and distribution of foodgrains. The arbitrarily fixed low prices increase demand but prove a disincentive for increased production. The farmer not only does not apply necessary inputs to secure larger production but diverts even the existing land, whenever he finds it profitable and possible to do so, to raising of non-food crops. This creates a perpetual gap between demand and supply which has to be bridged by imports. Though the author has examined the food problem and food imports at length, no attempt has been made to quantify the factors influencing either the food problem or the food imports. Further, the study relates to the period 1947 to 1968 only.

---

7. S.S. Madalgi, *Population and Food Supply in India*, Lalvani Publishing House, Bombay, 1970.
8. B.M. Bhatia, *India's Food Problem and Policy Since Independence*, Somaiya Publications Pvt. Ltd., Bombay, 1970.

V.M. Rao[9], in his study, takes a look at the food sector within Indian Agriculture, with a view to assessing its capacity to meet the increasing requirements of a growing population, which might cross the one billion mark when India steps into the twenty-first Century. He distinguishes four broad alternative states in which an economy may find itself in any given period as regards its food situation, *viz.*, 'famine', 'overt hunger', 'latent hunger' and 'adequate food'[10]. The broad effect of food policies in India since Independence has been to contain the recurring state of 'overt hunger', caused by erratic monsoons. India has begun to look beyond the 'overt hunger' state to come to grips with the core of its food problem, *viz.*, the 'latent hunger' state. The implication of this in terms of concrete policies to be pursued are investigated in his study by V.M. Rao. He concludes that, given the anticipated growth in the population even a modest target for improving the diets of the poorer sections in the population would need a large increase in the supply of food, roughly a four-fold increase by the end of the Century. According to him, with a proper set of policies in the decades ahead, India can certainly hope to find the solution to her food problem. Determined and conscientious efforts to increase the production of foodgrains, stricter measures of discipline in relation to the distribution of foodgrains and increasing and more effective coverage of the rural poor under the public distribution system would be the main objectives of these policies. However, the study has not gone into the food problem of India in relation to the foodgrain imports, in detail.

According to Ramachandran[11], contrary to the popular belief, the villain of the peace in India's food problem is not production lagging behind the population. The amazing thing is that the food problem

9. V.M. Rao, *Second India Studies: Food*, The MacMillian Company of India Ltd.,.Delhi, 1975.
10. 'Famine' indicates the state in which scarcity of food assumes catastrophic proportions to cause extensive loss of life. 'Overt hunger' is state when the situation of food is bad enough to cause hardships to even those groups in the population which, under normal circumstances get enough food or more. 'Latent hunger' is a state in which a sizeable section of the population suffers chronically from lack of food. 'Adequate food' is that happy state in which the problem of hunger is confined to a few persons on the fringes of society and is taken care of by the normal channels of individual and social charity. For a detailed discussion of these four states, see V.M. Rao, *op. cit.*, pp.8-9.
11. L. Ramachandran, *India's Food Problem : A New Approach*, Allied Publishers Pvt. Ltd., New Delhi. 1977.

continues in spite of the significant success in the field of agricultural production. The author tackles the food problem from an altogether new angle. Without making an assumption of improved technology or increased yields, he suggests a parallel approach, which aims at the maximum utilization of the increased production through the regulation of the pattern of production and consumption and through improved nutrition. However, the author has not examined food problem in relation to trends in foodgrain imports.

Badrul Islam[12] points out that the immense socio-economic and political influence of rice and its importance in international transactions has induced Governments in Asian rice deficit countries to intervene in foreign and domestic trade. Rice imports are a Government monopoly in almost all the Asian rice-importing countries, except Hong Kong, Malaysia and Singapore. Even in these countries, imports are regulated by a number of other measures. The author develops the hypothesis that significant proportion of Government interference in rice importing is motivated by a desire to conserve foreign reserves. He suggests that Asian rice surplus countries could benefit by importing non-rice products from the Asian rice deficit countries. However, the author has not made a detailed study of India, as his paper relates to six Asian rice importing countries, *viz.*, India, Korea, Malaysia, Pakistan, Philippines and Sri Lanka.

John Wall,[13] in his paper, reviews the elements and performance of the system of public foodgrain management in India. He discusses the arrangements made to run the system, including the policy formulation and the institutions which are required to implement it. He points out that the system helps in the promotion of the basic objectives of price support, price stability, avoidance of famine or nutritional stress. In this context, he examines the role of foodgrain imports also. However, the author has not examined the trends in foodgrain imports in detail as the emphasis was more on the overall management of foodgrains in India.

---

12. Badrul Islam, "Price, Income, and Foreign Exchange Reserve Elasticity for Asian Rice Imports", *American Journal of Agricultural Economics*, Vol. 60, No.3, August 1978, pp.531-535.
13. J. Wall "Foodgrain Management: Pricing, Procurement, Distribution, Import and Storage Policy", in World Bank, *India occasional Papers*, Staff Working Paper, No. 279, November 1978, pp. 43-72.

On the other hand, Shlomo Reutlinger,[14] in his paper, estimates the level of grain imports needed, averaging good and bad years together, using assumption about domestic production. He uses this average import level along with alternative hypothetical levels of storage capacity to hold bufferstocks in a quantitative model designed to analyse various import and bufferstocking strategies. The most attractive strategy is found to be a combination of imports and bufferstocking that stabilizes both domestic availability and the financial cost of grain imports. However, the analysis of the paper relates only to the period of 1953-54 to1975-76.

J.S. Sarma,[15] in his paper, analyes India's performance in terms of the achievement of self-sufficiency on the basis of past trends in foodgrain production and consumption. The question whether India will continue to be self-sufficient assumes great significance because according to the author the entry of India as a large buyer will have a major influence on the international market and will have large implications for World food situation. The author concludes that unless the growth in demand for foodgrain turns out to be substantially larger than expected, there is a high probability that India will be able to meet her foodgrain requirements from domestic production and will not have to go in for large-scale imports except in years of severe drought. The existence of large bufferstocks, according to the author, will enable the country to tide over temporary shortages arising from adverse weather conditions. However, the study covers the period 1950 to 1977 only.

Public distribution of food or rationing is an important facet of food management operations of a number of developing countries in the World. Rationing has major direct and indirect effects on various aspects of the national economy. Imports for rationing purposes restrict the foreign exchange that will be available for other sectors and other developmental activities. Thus, imports and local procurement for ration supply of foodgrains and other essential commodities have considerable influence on the level of economic

---

14. S. Reuglinger, "The Level of Stability of India's Foodgrain Consumption", in World Bank, *India : Occasional Papers*, *op. cit.*, 1978, pp. 92-224.
15. J.S. Sarma, "India - A Drive Towards Self-sufficiency in Foodgrains", *American Journal of Agricultural Economics*, Vol. 60, No.5, December 1978, pp.859-864.

growth of a country. There are a number of studies[16] on the impact of public distribution of foodgrains in India. Though these studies indicate the relation that exists between imports and public distribution, they have not examined the food problem and imports of foodgrains in detail.

Upadhyaya[17] has undertaken a study of import financing in India for the period 1969-78 . It covers the whole range of foreign exchange problems facing the economy and the investment strategy, import pattern, the method of paying for imports and related issues like import substitution and choice and development of techniques. He examines the trends in the import of foodgrains in the successive five-year plans and indicates the rising trend in the imports of foodgrains, both in terms of value and quantity, but has not examined the food problem faced by the Indian economy in detail.

S.K. Patel[18] in his study, provides for the first time an analytical-empirical research on Indian import behaviour both for the pre-devaluation and post-devaluation periods, based on quarterly observations. Imports have been disaggregated into different categories namely total imports, non-food imports, food imports, imports of raw-materials, imports of fuels, imports of chemicals, imports of fertilizers, imports of manufactures and imports of machinery and transport equipment. Import functions are formulated to ascertain the relative importance of various factors which influence imports. In examining the broad structural characteristics of imports of India, Patel examined the trends in food imports of India and found that the dependence on import of foodgrains has been a continuing feature of the Indian economy. However, Patel's study relates to the

---

16. Some of the studies are : P.S. George, *Public Distribution of Foodgrains in Kerala— Income Distribution Implications and Effectiveness*, International Food Policy Research Institute (IFPRI), Research Report-7, March 1979; J.S. sarma and Shyamal Roy and P.S. George, *Two Analyses of Indian Foodgrain Production and Consumption Data*, IFPRI, Research Report-12, November 1979; J.D. Gavan and I.S. Chandrasekera, *The Impact of Public Foodgrain Distribution on Food Consumption and Welfare in Sri Lanka*, IFPRI, Research Report-13, December 1979; and Harold Alderman and Joachim Von Braun, *The Effects of the Egyptian Food Ration and Subsidy System on Income Distribution and Consumption*, IFPRI, Research Report 45, July 1984.
17. K.K. Upadhyaya, *Import financing in India (With Special Reference to the Fourth and Fifth Five-Year Plans)*, Chugh Publications, Allahabad, 1980.
18. S.K. Patel, *Quarterly Econometric Analysis of Imports in India*, Hindustan publishing Corporation, Delhi, 1981.

period 1960-61 to 1972-73 only. It is also a broad study covering major categories of imports of India. Hence, an in-depth analysis of food imports of India in the context of the food problem of India has not been attempted by the author.

Adequate food intake by the poor is considered to be the key factor for better health and human well-being. In this context, increasing food production alone is not sufficient to ensure adequate food for all, but has to be combined with improved food distribution system. National efforts for enhanced food supply and improved food distribution systems have to be combined together with intercountry cooperative programmes to tackle the food problem. Food imports account for one-quarter of the value of imports of most Pacific Island countries. The high degree of dependence on imported foods could have nutritional and social effects, in times of decline in export prices or increase in import costs. These related problems have been examined by a special U.N. Committee on Agricultural Development.[19] However, the study has not examined food problem and imports of foodgrains in detail for India, as it is a relatively wider study relating to ten countries.

S.N. Kulshreshtha and D.D. Tiwari[20] reviewed the major problems and developments in agriculture in their paper. In this context, they also examined issues relating to food self-sufficiency, which has been the main objective since the beginning of the Five-Year Plans. One of the crude measures of food self-sufficiency is the substitution of food imports by domestic production. However, such an assessment of food self-sufficiency does not take into consideration institutional requirements and distributional problems in the availability of food due to differences in purchasing power. The authors maintained that although food self-sufficiency may have been attained in the Indian economy in a very narrow economic sense, it is far away from being achieved in terms of satisfying the nutritional requirements of the increasing masses of people in India. However,

---

19. U.N. ESCAP, *Committee on Agricultural Development, Special Study on Food Supply and Distribution Systems*, Bangkok, 12-18 January 1982. The Countries covered by the Report are : Bangladesh, China, India, Nepal, Pakistan, Sri Lanka, Indonesia, Malaysia, Philippines and the Republic of Korea.
20. S.N. Kulshreshtha and D.D. Tiwari, "Agriculture in India—Problems and Prospects", in J.S. Uppal (ed.) *India's Economic Problems : An Analytical Approach*, Tata McGraw-Hill Publishing Company Ltd., New Delhi, 1983, pp. 95-116.

the authors have not gone into the food problem and foodgrain imports in-depth in their study.

While a number of Third World countries have fallen into increasing food dependency, a remarkable aspect of Indian agricultural development is that it has allowed India to break away from chronic dependency on food imports and to attain self-sufficiency in foodgrains. However, food self-sufficiency has been reached without satisfying the food needs of the poorest 40 per cent of the population. The key to this problem according to Alain De Janvry and Subba Rao[21] lies in the level of food prices. In this context, the authors have discussed the trends in food production, consumption, stocks and imports since the amid-1960's when new seed varieties were introduced. In general, they found that imports declined in periods of favourable spurts in domestic production, while long-run foodgrain growth rates were matched by population growth rates. However, as the main purpose of their study has been to illuminate some of the growth and distributional effects of agricultural price policy in India, the authors have not examined the food problem or foodgrain imports of India in detail.

In the context of the federal structure of the Indian economy, the quantitative imbalance in the requirement of foodgrains at the regional level has assumed significance in recent years. Against this background, A.K. Mitra,[22] attempted a systematic study of the quantitative requirement of foodgrains in the regional perspective for the Indian economy mainly based on secondary data available from the Government sources. The author estimates the demand for and supply of foodgrains and shows the deviation between the two at the national as well as at the regional levels for the period 1951-52 to 2001-2002. He also examines the measures to tackle the imbalance between demand and supply. However, the author has not examined aspects relating to foodgrain imports in the context of India's food problem.

Lekhi[23] analysed the trends in agricultural development in India for the period 1947-48 to 1980-85. Pointing to the greater dependence

21. A. De Janvry and K. Subba Rao, *Agricultural Price Policy and Income Distribution in India*, Oxford University Press, Delhi, 1986.
22. A.K. Mitra, *Food Problem in India : A Quantitative Study in the Regional Perspective*, Anu Books, Meerut, 1989.
23. R.K. Lekhi, *Agricultural Development in India*, Classical Pub. Company, New Delhi, 1986, pp.1-6.

of the Indian economy on the vagaries of rainfall after partition, he examined briefly the trends in the imports of foodgrains of India during the period 1947-48 to 1951-52 and concluded that the financial strain of import of foodgrains had a menacing effect on the national economy. Though, the author made a painstaking study of the wide spectrum of agricultural policy and other development programmes aimed at building a self-sufficient and stable agrarian economy in the country. He has not gone into the food problem and food imports of India in detail.

Swant[24] has undertaken an assessment regarding the performance of Indian agriculture mainly with reference to two aspects, *viz.*, (*i*) pace of growth in the production of foodgrains and non-foodgrains for the country as a whole and for the major States and (*ii*) imbalances generated in the process of growth over the previous decades. For the period 1950-51 to 1983-84, the analysis makes it clear that the performance of Indian agriculture was quite impressive in several respects and has belied all the pessimistic presumptions both about food availability situation and about the prospects of growth in future. The author points out that the persistence of chronic poverty despite substantial growth in foodgrain production exphasises the need for maximum exploitation of the potential of India's natural resources. But, the author has not examined India's food problem in this paper at length.

Food imports to the developing world have been rising steadily during the last decade, as increase in the demand for food continues to exceed local production capabilities. This tendency seems likely to continue into the 1990s also. John Mellor,[25] explains how an era of global food adequacy might be brought about if a three-way partnership could be formed between the developed countries, the developing countries and agricultural research institutions. However, the author has not examined the position of India separately as the paper deals with World food situation.

24. S.D. Swant, "A Review of Performance in the Agricultural Sector", in P.R. Brahmananda and V.R. Panchamukhi (ed.), *The Development Process of the Indian Economy*, Himalaya Pub. House, Bombay, 1987, pp. 200-227.
25. J.W. Mellor, "Food and Development: The Critical Nexuz", *Economic Impact*, No. 61, May 1987, pp. 8-14.

Derek Byerlee[26] takes a new look at trends in food imports by developing countries and focuses on wheat imports in the context of institutional and external trade environment in which these imports. occur. A cross-country analysis of wheat imports and policy intervention is also undertaken. The author emphasises on the need to pay more attention to variables such as urbanization and consumer price policy in making projections in respect of 'food gap' and argues for more analysis for food policy at the country level in order to build up aggregate regional and world projections. However, an in-depth analysis of India's food problem and imports has not been attempted in this paper.

In his study, Paul Streeten[27] attempts to sort out the issues underlying the actual and apparent dilemmas confronted by policy-makers in the developing countries. The fundamental dilemma is between the need for high prices for food in order to increase agricultural production on the one hand and the need for low-prices to protect (at least in the short-run) the poor buyers of food. The secondary dilemma is that between the role of agricultural sector as a producer of food, which calls for its modernisation and its role as a provider of resources[28] for the rest of the economy, which calls for its exploitation. However, the author has not examined the problems relating to India in detail, as the study relates to developing countries broadly.

In his study, Gulati[29] aims at empirically mapping the various considerations that go into the fixation of support/procurement price, its effects on its own free market price, issue price, quantum of wheat procurement etc. The purpose of the study is to identify and empirically estimate the various interlinkages that characterise the wheat sector and also to work out the effects of certain policy packages. In this context, the author has also undertaken an analytical study of wheat offtake from PDS, as well as wheat imports from

26. D. Byerlee, "The Political Economy of the Third World Food Imports : The Case of Wheat", *Economic Development and Cultural Change*, Vol. 35, No.2, Jan. 1987, pp. 307-29.
27. P. Streeten, *What Price Food? Agricultural Price Policies in Developing Countries*, St. Martin's Press, New York, 1987.
28. The resources may be savings for industrialisation or foreign exchange or tax-revenue.
29. A. Gulati, *Agricultural Price Policy in Indi : An Econometric Approach (with special reference to wheat)*, Concept publishing Company, New Delhi, 1987.

abroad. However, the author has not examined foodgrain imports in detail.

Ramesh Babu[30] in his study, analyses the trends in the composition of India's imports and estimates the determinants of imports at the aggregate level as well as for selected products. The analysis indicates a decline in the shares of both food and non-food imports. Consumer good imports appear to have been influenced mostly by food imports, which witnessed volatile fluctuations during the period of study. The study, however, has not examined food imports in detail against the background of food problem in India. Further, it relates only to the period 1958-1977.

In his survey of India's rural development, Gilbert Etienne[31] examines foodgrain statistics and population censuses which point to improvement in the overall food situation. Most Asian countries like India, China, Indonesia, Pakistan and Philippines have succeeded in raising foodgrians production so that imports have fallen or disappeared. They have achieved this through the progress of irrigation, new seeds and chemical fertilizers, and the pursuit of a policy of grain procurement and storage. The author maintains that achievement of self-sufficiency in cereal production should not lead to complacency and relaxation on the food front. However, the author has not examined food problem in India in detail in relation to foodgrain imports.

2.2. Limitations of the Earlier Studies and the Need for the Present Study

The study of food problem in India received considerable attention of the academicians as well as public authorities, after the Bengal Famine in 1943 and also since Independence. The deficiency in food supply in India in relation to demand became chronic after the Second World War and the partition of the country. The rapid rise in population and purchasing power in the hands of the people led to increase in the demand for food, which could not be matched by the production of foodgrains, resulting in rise in foodgrain prices and frequent occurrances of periods of scarcity. Various structural,

30. S. Ramesh Babu, *India's Foreign Trade : Some Trends*, Chugh Publications, Allahabad, 1988.
31. G. Etienne, *Food and Poverty : India's Half Won Battle*, Sage Publications, New Delhi, 1988.

technological and economic measures have been adopted to bring about adjustments between supply and demand for foodgrains. These measures have resulted in the easing of the problems on the food front to some extent in the eighties.

Some of the earlier studies on foodgrains confined themselves to selected issues connected with either production or procurement or distribution or demand aspects of foodgrains. The level of production of foodgrains is an important determinant of the level of imports of foodgrains. As the domestic production of foodgrains increases, the level of imports tends to decline. But production itself depends upon factors such as area, yield and profitability. Hence, there is need for a systematic analysis of various aspects of foodgrain imports of India. The present study attempts to bring together the food problem and foodgrain imports of India. It attempts to study quantitatively foodgrain economy as well as foodgrain imports of India.

Further, the earlier studies confined themselves to a study of the RITC Section 'O' relating to Food and Live Animals group as a whole. No attempt has been made to study separately the trends in the imports of foodgrains. The studies also related to early periods. In the present study, an attempt is made to examine the trends in the production of foodgrains along with the trends and determinants of foodgrain imports of India separately.

Thus, in the present study an attempt is made to deal with a particular aspect of the problem, *i.e.*, the study of foodgrain economy and foodgrain imports of India. On the basis of current trends, efforts are made to identify the role of food imports in the Indian economy in the context of the food problem for the period 1955-56 to 1986-87, for which comparable data are available. The present study is thus an attempt to fill the gap in the literature to some extent.

2.3. Objectives of the Study

The specific objectives of the study are as follows :

(*a*) to analyse the growth of the foodgrain economy of India in terms of area, production and yield per hectare of foodgrains for the country as a whole;

(*b*) to study the trends in the prices, per capita net availability, procurement, public distribution and stocks of foodgrains as well as targets and shortfalls in foodgrain production in the various Five-Year Plans of India;

(*c*) to examine the trends in the imports of foodgrains of India, the pattern and direction of such imports and the import-availability ratio for foodgrains; and

(*d*) to identify the factor s which determine the demand for imports of foodgrains of India and to suggest policy measures.

2.4. Methodology of the Study

In order to examine the above objectives, the following techniques have been employed in the present study.

(*a*) *Trend Anlysis*

In the study of the growth of the foodgrain economy, tabular analysis is employed to analyse the trends in area, production and yield per hectare of foodgrains for the country as a whole, for the period 1955-56 to 1986-87. Growth rates are estimated by fitting an exponential function of the following form :

$$Y = ae^{bt}$$

where,

Y = Area/Production/Yield of foodgrains

a = constant

b = growth rate

t = period of time

Growth rates are estimated to indicate growth in the area, production and yield of foodgrains. Growth rates are estimated for the total period 1955-56 to 1986-87, as well as for the sub-periods namely, 1955-56 to 1966-67, representing the pre-green Revolution period and 1967-68 to 1986-87, representing the post-green Revolution period.

(*b*) *Coefficient of Variation*

The present study uses coefficient of variation for analysing the instability in area, production and yield of foodgrains in the Indian economy by employing the following formula :

$$C.V. = \sigma x\, 100$$

where,

$C.V.$ = coefficient of variation

σ = standard deviation

$\bar{x}$ = mean

The analysis of area, production and yield of foodgrains of India *vis-a-vis* other developing countries, developed countries and the World as a whole, changes in cropping pattern as well as production pattern of foodgrains of India, structure of imports of India, food and non-food imports and pattern of foodgrain imports are analysed with the help of percentage shares and three-year moving averages. The moving average method is used to overcome any possible short-term cyclical fluctuations.

Exponential growth rates and coefficients of variation are computed for prices, per capita net availability, procurement, stocks and distribution of foodgrains. Growth rates are also estimated for imports of foodgrains, imports of Food and Live Animals, chiefly for Food and the total imports of India.

(*c*) *Regression Analysis*

To analyse the factors determining demand for imports of foodgrains of India, the study employs multiple regression analysis for the following function namely,

$$Qmf_t = F(PY_t, RP_{ft}, P_{ft}, S_{f(t-1)}, PR_{ft}, EA_t, NFA_t)$$

Where,

Qmf_t = Quantum of imports of foodgrains of India, in year 't'

PY_t = Per capita income of India in year 't'

RP_{ft} = Relative price Index of foodgrains in year 't'

P_{ft} = Production of foodgrains in year 't'

$S_{f(t-1)}$ = Stocks of foodgrains in year 't-1'

PR_{ft} = procurement of foodgrains in year 't'

EA_t = external assistance in year 't'

NFA_t = net foreign assets in year 't'

t = time period

In addition to the above set of explanatory variables, dummy variables to capture the effects of policy changes relating to maintenance of bufferstocks and procurement of foodgrains that took place in 1975-76 have been added. The above function is estimated for the period 1955-56 to 1986-87, because of the availability of comparable data for all the variables involved in order to carry out the regression analysis for foodgrain imports of India.

2.5. Source of data

Data relating to area, production and yield of foodgrains are obtained from various issues of 'Estimates of Area and Production of Principal Crops in India', and 'Economic Survey', published by Government of India. Wholesale prices of Foodgrains are obtained from various issues of 'Report on Currency and Finance', published by Reserve Bank of India. Procurement, public distribution and stocks of foodgrains are obtained from various issues of 'Bulletin on Food Statistics', published by Ministry of Agriculture, Government of India.

Data relating to value of imports of Cereals and Cereal preparations, imports of Food and live Animals chiefly, for Food, total imports of India, data relating to quantity of imports of foodgrains and import prices of foodgrains are obtained from various issues of 'Bulletin on Food Statistics,' published by Government of India, 'Trade Year Book,' published by FAO, 'Year book of International Trade Statistics,' published by United Nations and 'Monthly Statistics of Foreign Trade of India,' published by DGCIS, Calcutta. Data relating to per capita income, external assistance and net foreign asset are obtained from various issue of 'Economic Survey,' published by Government of India.

2.6. Plan of the Study

The study is organised into six Chapters. Chapter one examines the nature of food problem in India and brings out the importance of foodgrains and foodgrain imports in the Indian economy. A review of the earlier studies on food problem in general and on foodgrain imports in particular and the need for the present study are explained in Chapter Two. The objectives of the present study as well as the various sources of the data used in the study are also presented in this Chapter. Chapter Three deals with the growth of the foodgrain economy of India. An attempt is made in this Chapter to analyse the trends in the area, production and yield of foodgrains in the Indian economy in the Pre-Green Revolution and Post-green Revolution periods. Trends in prices, per capita net availability, procurement, public distribution and stocks of foodgrains, targets and shortfalls in foodgrain production in India are also examined in this Chapter. Chapter Four examines trends in the imports of foodgrains of India in terms of value, Volume and direction. The degree of dependence of the economy on imports of foodgrains is also analysed in this Chapter. Chapter Five examines the factors which determine the demand for imports of foodgrains and their relative importance. A brief summary and conclusions of the study are given in the Sixth Chapter.

CHAPTER 3

FOODGRAIN ECONOMY OF INDIA

3.1. Introduction

IN the Indian economy, a relatively greater attention is paid to foodgrains mainly because their production has fallen far short of demand during most of the past thirty years and also as these crops constitute the major dietary intake of the people of the Indian economy.[1] In view of the importance of foodgrains to the Indian economy, an in-depth analysis of foodgrain economy of India is taken up here. In particular, the present Chapter analyses: (*a*) trends in the area, production and yield of foodgrains in India relative to those of the World; (*b*) the cropping and production pattern of foodgrains; (*c*) trends in the wholesale prices of foodgrains; (*d*) per capita net availability of foodgrains; (*e*) procurement, public distribution and stocks of foodgrains and (*f*) targets and shortfalls in foodgrain production.

3.2. World Production of Cereals

Most countries produce a variety of staples. The predominant food various from area to area, but it is supplemented by other cereals or roots to some extent. Cereals far outweigh all other foodstuffs as sources of food energy, and in many developing countries contribute a major part of the protein as well. U.N., F.A.O., estimated that cereals, starchy roots and sugar provide 70 per cent in the four developing regions *viz.*, (*a*) Far East, (*b*) Near East, (*c*) Africa and (*d*) Latin America. Among the cereals, rice and wheat from the staple diet

---

1. F.K. Wadia, "Agricultural Price Policy", in J.N. Mongia (ed.), *India's Economic Policies 1947-77*, Allied Publishers Pvt. Ltd., New Delhi, 1980, p.150.

for probably 4 out of 5 of the World's population. Although rice is usually the staple food in the Far East, other grains and wheat predominate in large parts of India and Pakistan.[2]

Throughout the Far East, the most preferred food is rice. The Indian sub-continent absorbs approximately one-half of the starchy staples of the Far East excluding China (Mainland) and is easily the heaviest single consumer of cereals. Trends in the production of wheat, paddy, coarse cereals and total cereals of the World are shown in Table 3.1. and Diagram 3.1. It is evident from Table and the Diagram that World production of wheat (excluding USSR) increased from 146.9 million tonnes in 1950 to 522.9 million tonnes in 1984. Similarly, World production of paddy increased from 152.9 million tonnes in 1950 to 471.2 million tonnes in 1984. World production of cereals increased from 294.0 million tonnes in 1950 to 812.0 million tonnes in 1984 while World production of total cereals increased from 593.9 million tonnes in 1950 to 1,806.1 million tonnes in 1984.

Table 3.1 : Production of Cereals in the World

(Million Tonnes)

Year	Wheat	Paddy	Coarse Cereals	Total Cereals
1950[a]	146.9	152.9	294.0	593.9
	(46.1)	(48.1)	(61.2)	(53.3)
1960	243.7	239.5	486.4	969.6
	(76.5)	(75.8)	(101.2)	(86.9)
1970	318.4	316.1	480.5	1,115.0
	(100.0)	(100.0)	(100.0)	(100.0)
1971	353.9	317.6	434.7	1,206.2
	(111.1)	(100.0)	(90.5)	(108.2)
1972	348.3	308.0	616.4	1,272.7
	(109.4)	(97.4)	(128.3)	(114.1)
1973	376.2	336.0	662.5	1,374.7
	(118.2)	(106.3)	(137.9)	(123.3)
1974	364.2	33.0	641.3	1,338.5
	(114.4)	(105.3)	(133.5)	(120.0)
1975	360.2	358.7	652.7	1,371.6
	(113.7)	(113.5)	(135.8)	(123.0)

(Contd.)

2. U.N., F.A.O., Commodity Bulletin Series, *The Economic Relationships Between Grains and Rice*, Report-39, USA, 1965, p.7.

Table 3.1 (Contd.)

Year	Wheat	Paddy	Coarse Cereals	Total Cereals
1976	425.7	350.4	704.8	1,480.9
	(133.7)	(110.9)	(146.7)	(132.8)
1977	387.6	371.5	712.4	1,471.5
	(121.7)	(117.5)	(148.3)	(131.9)
1978	451.3	387.3	764.0	1,602.6
	(141.7)	(122.5)	(159.0)	(143.7)
1979	429.3	377.0	749.2	1,555.0
	(134.8)	(119.3)	(155.9)	(139.5)
1980	446.5	399.2	720.2	1,566.0
	(140.2)	(126.2)	(149.9)	(140.4)
1981	453.8	411.8	788.0	1,653.6
	(142.5)	(130.3)	(163.9)	(148.3)
1982	485.0	422.7	792.6	1,700.3
	(152.3)	(133.7)	(164.9)	(152.5)
1983	497.0	449.9	694.4	1,641.2
	(156.1)	(142.3)	(144.5)	(147.2)
1984	522.9	471.2	812.0	1,806.1
	(164.2)	(149.1)	(168.9)	(161.9)

Notes : (a) Excludes data for USSR.

Figures in Brackets represent index numbers with base 1970 = 100.

Source : GOI, Ministry of Agriculture, *Bulletin on Food Statistics*, 1985, p.233.

3.2.1. *Area, Production and Average Yield per Hectare of total Cereals in Major Producing Countries of the World :*

Area, production and average yield per hectare of total cereals in the major producing countries of the World are shown in Table 3.2. It is evident from the Table that Asia ranks first in terms of area and production of total cereals among the major producing Groups of the World in 1984. Area under total cereals in Asia increased from 300.9 million hectares in 1974-76 to 312.1 million hectares in 1984, while the production of total cereals in Asia increased from 548.3 million tonnes to 776.1 million tonnes during the same period. Yield per hectare of total cereals of Asia increased from 1,822 kg to 2,487 kg during this period. Yield per hectare of total cereals is found to be highest in respect of France (5.977 kg) among the major cereal producing countries of the World followed by USA (4,378 kg) and China (3,894 kg) in 1984.

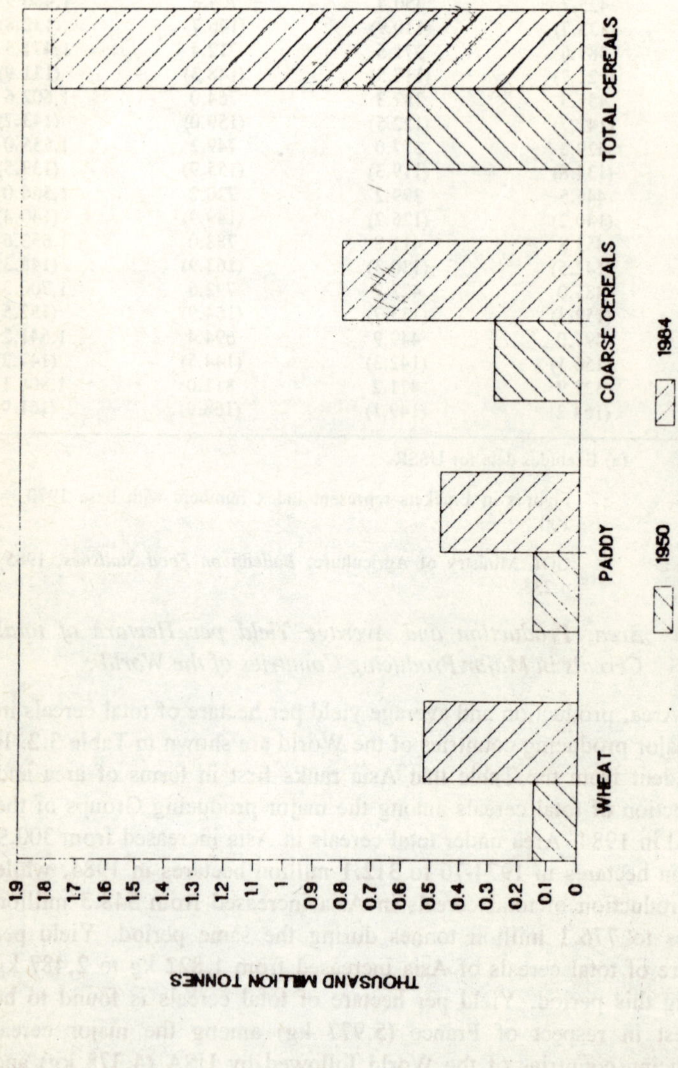

Diagram 3.1. Production of Wheat, Paddy, Coarse Cereals and Total Cereals in the World

India accounted for 6.67 per cent of the area under total cereals of the World and 10.67 per cent of total production of cereals of the World in 1984. From Table 3.2, it is evident that while the area under total cereals in India is relatively high, the yield per hectare is relatively low compared to the other major cereal producing countries of the World, (except USSR). The yield of total cereals in India of 1,545 kg per hectare in 1984 is lower than 2,472 kg of the World as well. Thus, cereal production in India is faced with the problem of relatively low yield per hectare.

Table 3.2 : Area, Production and Average yield per Hectare of Total Cereals@ in Major Producing Countries of the World

A—Area in Million Hectares
P—Production in Million tonnes
Y—Yield per hectare in Kgs.

Name of the Group/ Country		1974-76*	1981	1982	1983	1984
	(1)	(2)	(3)	(4)	(5)	(6)
Africa	A	69.5	73.0	71.6	70.8	70.9
	P	69.9	79.0	73.7	63.3	63.9
	Y	1,006	1,082	1,030	895	901
North Central America	A	101.9	114.0	109.2	95.7	108.8
	P	295.2	415.1	412.0	282.6	389.3
	Y	2,895	3,641	3,774	2,954	3,577
Canada	A	18.5	21.6	21.5	21.8	22.2
	P	37.6	50.8	53.5	47.6	43.2
	Y	2,027	2,348	2,386	2,187	2,035
U.S.A	A	71.1	78.0	76.2	58.6	71.8
	P	237.5	333.8	333.3	207.4	314.4
	Y	3,339	4,278	4,374	3,539	4,378
South America	A	37.5	38.7	41.7	36.8	37.6
	P	61.6	74.0	80.1	70.9	76.8
	Y	1,641	1,913	1,922	1,928	2,043
Brazil	A	19.6	19.9	21.9	18.1	19.7
	P	27.9	32.1	33.9	29.2	32.6
	Y	1,420	1,613	1,548	1,615	1,665
Asia	A	300.9	306.0	298.1	308.0	312.1
	P	548.3	660.5	676.6	748.4	776.1
	Y	1,822	2,159	2,270	2,430	2,487

(Contd.)

Table 3.2 (Contd.)

(1)		(2)	(3)	(4)	(5)	(6)
China	A	98.1	92.7	91.0	93.2	93.7
	P	243.2	286.5	314.8	345.8	364.8
	Y	2,479	3,092	3,458	3,710	3,894
India	A	100.7	105.5	101.1	106.2	109.6
	P	118.7	147.6	136.1	166.4	169.3
	Y	1,179	1,399	1,346	1,567	1,545
Indonesia	A	10.9	12.3	11.1	12.1	12.7
	P	25.6	37.3	36.8	40.3	43.3
	Y	2,338	3,020	3,331	3,326	3,408
Europe	A	71.0	69.9	70.1	69.4	69.8
	P	225.6	246.1	270.7	258.4	307.8
	Y	3,178	3,521	3,862	3,724	4,411
France	A	9.7	9.7	9.7	9.4	9.7
	P	36.5	45.0	48.7	46.5	57.8
	Y	3,776	4,614	5,003	4,964	5,977
Oceania	A	12.6	17.1	16.4	19.2	18.4
	P	18.0	24.5	15.1	32.1	29.7
	Y	1,420	1,432	918	1,677	1,609
Australia	A	12.4	16.9	16.2	18.9	18.2
	P	17.2	23.7	14.2	31.2	28.4
	Y	1,384	1,399	873	1,646	1,563
USSR	A	122.1	119.9	117.9	116.2	112.9
	P	179.0	154.4	172.0	185.5	162.5
	Y	1,466	1,288	1,459	1,596	1,439
World	A	715.6	738.6	724.9	715.9	730.6
	P	1,397.5	1,653.6	1,700.3	1,641.2	1,806.1
	Y	1,953	2,239	2,346	2,293	2,472

Note : @ Total cereals include rice in terms of paddy.
* 1974, 1975 and 1976 average.

Source: 1. FAO, *Production Year Book*, 1983.
2. GOI, *Bulletin in Food Statistics*, 1985.

3.3. Trends in the Area, Production and Yield of Foodgrains in India

Trends in the area, production and yield of foodgrains in the Indian economy during the period 1955-56 to 1986-87 are shown in Table 3.3. Exponential growth rates are estimated for area., production and yield of foodgrains for the period 1955-56 to 1986-87. The most striking development in 1965-66 is the introduction of the New Development Strategy for Agriculture, popularly known as the Green Revolution. Hence, growth rates are also estimated for the period 1955-56 to 1966-67, indicating the Pre-Green Revolution period and for 1967-68 to 1986-87 indicating the post-Green Revolution period. Growth rates are estimating by using the formula given in Chapter 2, as follows :

$$Y = ae^{bt}$$

where,

Y = area/production/yield of foodgrains

a = constant

b = growth rate and

t = time period

Coefficient of variation is obtained for area, production and yield for the same periods. The coefficients of variation and the estimated growth rates are given in Table 3.4. Trends in the area under foodgrain cultivation are analysed below.

3.3.1. *Trends in the Area under Foodgrains in India*

Trends in the area under foodgrains in India are shown in column (2) of Table 3.3 and Graph 3.1. The coefficient of variation and the estimated growth rates for area are shown in Table 3.4. It is evident from Tables 3.3 and 3.4. that area under foodgrains in India increased from 110.56 million hectare in 1955-56 to 115.30 million hectares in 1966-67, *i.e.*, at a rate of 0.52 per cent per annum during the pre-Green Revolution period. During the post-Green Revolution period, area under foodgrains increased from 121.42 million hectares

0.31 per cent per annum. Thus, the growth rate of area under foodgrains is found to be less in the post-Green Revolution period compared to that during the pre-Green Revolution period.

On the whole, area under foodgrains in India increased from 110.56 million hectares in 1955-56 to 127.20 million hectares in 1986-87, *i.e.*, at a rate of 0.49 per cent per annum. It is evident that there has not been much expansion in the area under foodgrains in India during the period 1955-56 to 1986-87.

Table 3.3 : Area, Production and Yield of Foodgrains in India

Area in Million hectares
Production in Million tonnes
Yield per hectare in kg.

Year	Area	Production	Yield
(1)	(2)	(3)	(4)
1955-56	110.56	69.3	605
1956-57	111.14	72.4	629
1957-58	109.48	66.6	587
1958-59	114.76	78.8	672
1959-60	115.82	77.1	662
1960-61	115.58	82.3	710
1961-62	117.23	82.7	705
1962-63	117.84	80.2	680
1963-64	117.42	80.6	687
1964-65	118.11	89.4	757
1965-66	115.10	72.3	629
1966-67	115.30	74.2	644
1967-68	121.42	95.1	783
1968-69	120.43	94.0	781
1969-70	123.57	99.5	805
1970-71	124.32	108.4	872
1971-72	122.62	105.1	858
1972-73	119.28	97.0	813
1973-74	126.54	104.7	827
1974-75	121.08	99.8	824
1975-76	128.18	121.0	944
1976-77	124.36	111.2	894
1977-78	127.51	126.4	991
1978-79	129.01	131.9	1,022
1979-80	125.21	109.7	876
1980-81	126.67	129.6	1,023
1981-82	129.14	133.3	1,032
1982-83	125.09	129.5	1,035
1983-84	131.16	152.4	1,162
1984-85	126.67	145.5	1,149
1985-86	128.02	150.4	1,175
1986-87	127.20	143.4	1,128

Source : 1. GOI, Ministry of Food and Agriculture, *Estimates of Area and Production of Principal Crops in India*, New Delhi, various issues
2. GOI, Ministry of Finance, *Economic Survey 1988-89*, Delhi.

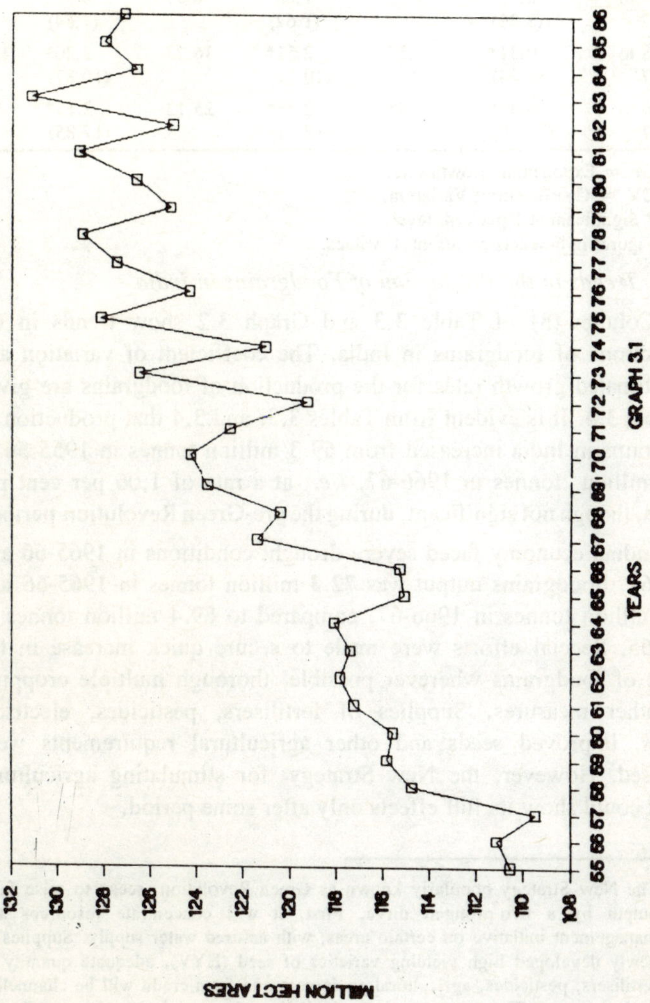

Graph 3.1. Trends in the Area under Foodgrains in India

Table 3.4 : Growth Rates of Area, Production and Yield of Foodgrains in India

Period	Area		Production		Yield	
	Gr.	CV	Gr.	CV	Gr.	CV
1955-56 to 1966-67	0.52* (3.36)	2.55	1.06 (1.62)	8.37	0.83 (1.64)	7.26
1967-68 to 1986-87	0.31* (4.34)	2.57	2.51* (10.14)	16.23	2.20* (10.87)	14.18
1955-56 to 1986-87	0.49* (12.56)	4.95	2.55* (17.71)	25.12	2.13* (17.85)	2.11

Note: Gr. = Exponential growth rate.
CV = Coefficient of Variation.
* Significant at 1 per cent level.
Figures in brackets represent 't' values.

3.3.2. *Trends in the Production of Foodgrains in India*

Column (3) of Table 3.3 and Graph 3.2 show trends in the production of foodgrains in India. The coefficient of variation and the estimated growth rates for the production of foodgrains are given in Table 3.4. It is evident from Tables 3.3. and 3.4 that production of foodgrains in India increased from 69.3 million tonnes in 1955.56 to 74.2 million tonnes in 1966-67, *i.e.*, at a rate of 1,06 per cent per annum, though not significant, during the pre-Green Revolution period.

Indian economy faced severe drought conditions in 1965-66 and 1966-67. Foodgrains output was 72.3 million tonnes in 1965-66 and 74.2 million tonnes in 1966-67, compared to 89.4 million tonnes in 1964-65. Special efforts were made to secure quick increase in the output of foodgrains wherever possible, thorough multiple cropping and other measures. Supplies of fertilisers, pesticides, electrical motors, improved seeds and other agricultural requirements were increased. However, the New Strategy[3] for stimulating agricultural output could show its full effects only after some period.

---

3. The New Strategy popularly known as Green Revolution, seeks to raise farm output by a two-pronged dirve. First, it will concentrate resources and management initiative on certain areas, with assured water supply. Supplies of newly developed high yielding varieties of seed (HYV), adequate quantity of fertilisers, pesticides, agricultural equipment and farm credit will be channelled to these areas. The second aspect of the New Strategy is to raise rapidly the proportion of irrigated area which is double cropped by the introduction of quickly naturing seed varieties of cereals. Taken together, the two programmes which comprise the New Agricultural Strategy are aimed at obtaining a substantial increase in the total foodgrain output. See GOI, *Economic Survey* 1966-67, Delhi, pp. 11-12.

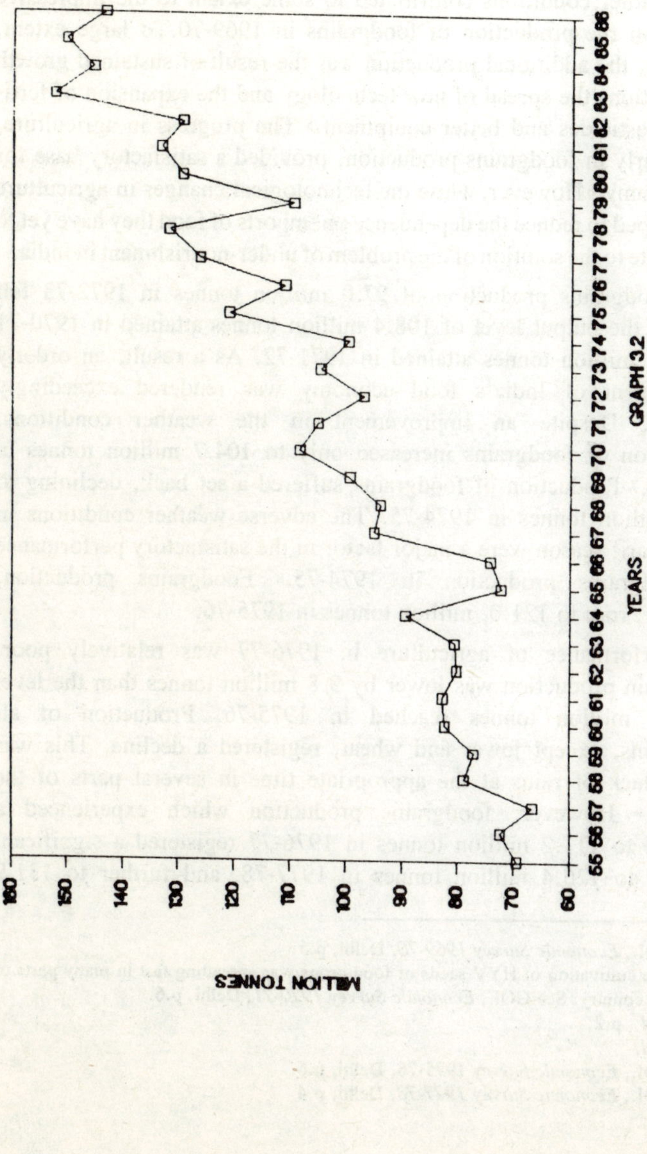

GRAPH 3.2

Graph 3.2 : Trends in the Production of Foodgrains in India

Foodgrains production, which had increased to 95.1 million tonnes in 1967-68 was slightly lower at 94.0 million tonnes in 1968-69. The decline in foodgrains production in 1968-69 was due to a fall in the production of coarse-grains and pulses.[4] In 1969-70, however, foodgrains production once again rose to 99.5 million tonnes. Favourable weather conditions contributed to some extent to the impressive growth in the production of foodgrains in 1969-70.To large extent, however, the additional production was the result of sustained growth in irrigation, the spread of new technology and the expansion of fertilisers, pesticides and better equipment.[5] The progress in agriculture, particularly in foodgrains production, provided a satisfactory base for the economy.[6] However, while the technological changes in agriculture have helped to reduce the dependence on imports of food they have yet to contribute to the solution of the problem of under-nourishment in India.

Foodgrains production of 97.0 million tonnes in 1972-73 fell short of the output level of 108.4 million tonnes attained in 1970-71 and 105 million tonnes attained in 1971-72. As a result, an orderly management of India's food economy was rendered exceedingly difficult. Despite an improvement in the weather conditions, production of foodgrains increased only to 104.7 million tonnes in 1973-74.[7] Production of foodgrains suffered a set back, declining to 99.8 million tonnes in 1974-75. The adverse weather conditions in 1974 kharif season were a major factor in the satisfactory performance of foodgrains production in 1974-75.[8] Foodgrains production, however, rose to 121.0, million tonnes in 1975-76.

Performance of agriculture in 1976-77 was relatively poor. Foodgrain production was lower by 9.8 million tonnes than the level of 121 million tonnes reached in 1975-76. Production of all foodgrains, except jowar and wheat, registered a decline. This was due to lack of rains at the appropriate time in several parts of the country.[9] However, foodgrains production which experienced a deadline to 111.2 million tonnes in 1976-77 registered a significant increase to 126.4 million tonnes in 1977-78, and further to 131.9

---

4. GOI., *Economic Survey 1969-70*, Delhi, p.5.
5. The cultivation of HYV seeds of food crops was spreading fast in many parts of the country. See GOI., *Economic Survey 1970-71*, Delhi, p.6.
6. *Ibid.*, p.2.
7. *Ibid.*
8. GOI., *Economic Survey* 1975-76, Delhi, p.6.
9. GOI., *Economic Survey 1977-78*, Delhi, p.4

million tonnes in 1978-79, but suffered a set back and declined to 109.7 million tonnes in 1979-80.[10]

Increased foodgrain production of 129.6 million tonnes was achieved in 1980-81, despite erratic rainfall in several parts of the country. As the cropped area increased only marginally, the increase in production was primarily due to increase in yield per hectare,[11] as can be seen from Col. (4) of Table 3.3.

Foodgrains production increased from 129.6 million tonnes in 1980-91 to 133.3 million tonnes in 1981-82. However, the two years, *viz.*, 1982-83 and 1983-84 witnessed extremely unusual weather conditions. The 1982-83 drought conditions brought foodgrain output below trend and the exceptionally good monsoon in 1983-84 took the growth rate far above the long-term trend.[12] Foodgrains production reached a peak level of 152.4 million tonnes in 1983-84, but declined to 145.5 million tonnes in 1984-85, due to unfavourable rainfall.[13]

The first three years of the Seventh Five-Year Plan were poor monsoon years. Consequently, 1985-86 witnessed a production level of 150.4 million tonnes of foodgrains, while the second year 1986-87, recorded a decline in production to 143.4 million tonnes. In 1987-88 also there was a substantial decline in production of foodgrains to 138.4 million tonnes.[14] The momentum of foodgrains growth established during the early eighties, leading upto a peak level production of foodgrains in 1983-84, has been seriously disturbed by adverse weather conditions during the last few years.[15]

---

10. The Decline in output of this order adversely affected all sectors of the economy. However, the resilience that the economy acquired is evident from the fact that whereas in the wake of comparable drought of 1965-66 and 1966-67, the country had to import about 7.46 million tonnes and 10.36 million tonnes of foodgrains respectively, it was possible during the 1979-80 drought to meet the requirements of public distribution system and other programmes entirely from the domestic buffer stocks built up in the earlier years. See GOI., *Economic Survey 1981-82*, Delhi, p.6
11. GOI., *Economic Survey 1982-83*, Delhi, p.5.
12. GOI., *Economic Survey 1985-86*, Delhi, p.7
13. *Ibid.*, p.6.
14. GOI., *Economic Survey 1988-89*, Delhi, p.10
15. GOI., *Economic Survey 1987-88*, pp.6-8. However, large stocks, built up to over 23 million tonnes by June 1987, made it possible to maintain adequate supplies of foodgrains, especially in the drought affected areas through the public distribution system. This programme of maintaining adequate release of foodgrains has formed the back bone of Government's policy in combating the drought. See GOI., *Economic Survey 1987-88*, pp. 1-3.

Thus since 1967-68, production of foodgrains showed an increasing trend, but with fluctuations. Production of foodgrains increased from 95.1 million tonnes in 1967-68 to 143.4 million tonnes in 1986-87, *i.e.*, at a rate of 2.51 per cent per annum, during the post-Green Revolution period.

On the whole, production of foodgrains in India increased from 69.3 million tonnes in 1955-56 to 143.4 million tonnes in 1986-87, *i.e.*, at a rate of 2.55 per cent per annum. Thus, there is an increase in the production of foodgrains but with considerable fluctuations as can be seen from Graph 3.2. The peak level of 152.4 million tonnes of production of foodgrains was reached in 1983-84. Thereafter, foodgrains production showed a decline to 143.4 million tonnes in 1986-87, the year in which the economy faced severe drought conditions. Poor monsoons for three consecutive years culminated in one of the worst drought conditions on record in 1987.[16]

Like other countries of the World, large scale use of modern agricultural technology is an important variable responsible for the present stage of foodgrains production in India. The variation in the production of foodgrains during the period largely owes to better utilisation of the existing resources and favourable weather conditions. The middle of the sixties can be treated as a turning point in India in Indian agriculture, called as 'Green Revolution'. Large scale use of technological factors like chemical fertilisers, HYV seeds, irrigation, mechanical implements and pesticides got due emphasis only since the severe drought of 1965-66. After the introduction of High Yielding Varieties Programme (HYVP) during 1966-67, there has been a sharp increase in the production of foodgrains.

Following the recommendations made in the course of the Mid-Term Appraisal of the Seventh Plan, a Task Force was set up to prepare an Action Plan to achieve the foodgrains production targets of the Seventh Plan. The principal strategy in the Special Foodgrains Production Programme (SFPP) is to capitalise on identified sources of immediate agricultural growth.[17]

---

16. However, as a result of exceptionally good monsoon combined with the thrust programmes like Special Foodgrains Production Programme (SFPP) and Special Rice Production Programme (SRPP), foodgrains production touched a record level of 170.3 million tonnes in 1988-89 as against 140.4 million tonnes in 1987-88. GOI., *Economic Survey 1989-90*, Delhi, p. 10.
17. GOI., *Economic Survey 1988-89*, p.10

3.3.3. Trends in the Yield per Hectare of Foodgrains in India

Trends in the yield per hectare of foodgrains in India are shown in Col.(4) of Table 3.3. and Graph 3.3. The coefficient of variation and the estimated growth rates are given in Table 3.4. It is evident from Tables 3.3. and 3.4 that the yield per hectare of foodgrains in India increased from 605 kg in 1955-56 to 644 kg in 1966-67, *i.e.*, at a rate of 0.93 per cent per annum, though not significant, during the Pre-Green Revolution period. Yield per hectare started showing significant improvements during the post-Green Revolution period. It increased sharply from 783 kg in 1966-67 to 1.128 kg in 1986-87, *i.e.*, at a rate of 2.20 per cent per annum, during the Post-Green Revolution period.

On the whole, yield per hectare of foodgrains in India increased from 605 kg in 1955-56 to 1.128 kg in 1986-87, *i.e.*, at a rate of 2.13 per cent per annum. This trend in the yield per hectare of foodgrains in India can also be seen from Graph 3.3. The peak level of 1,175 kg per hectare is reached in the year 1985-86, While it is interesting to note that there is a substantial increase in the yield per hectare due to the implementation of New Agricultural Strategy, the yield per hectare in India is still found to be significantly lower than that of the developed countries like USA and UK. In fact, the crop yield in India continues to be one of the lowest in the World. The greatest single limiting factor contributing to this low yield per hectare is the uncertain availability of water.[18]

Thus, it is evident from Table 3.4 that the production of foodgrains in India increased at a rate of 2.55 per cent annum. While the area under foodgrains increased at a rate of 0.49 per cent per annum and yield per hectare increased at a rate of 2.13 per cent per annum, during the period 1955-56 to 1986-87. From this it may be concluded that increase in the production of foodgrains was mostly due to increase in yield per hectare. However, during the pre-Green Revolution period, *i.e.*, 1955-56 to 1966-67, increase in production of foodgrains, though not significant, was mostly due to increase in area under foodgrains (at a rate of 0.52 per cent per annum). During the post-Green Revolution period, *i.e.*, 1967-68 to 1986-87, increase in production of foodgrains was mostly due to increase in yield per

18. V.G. Panse, "Trends in Area and Yield of Principal Crops in India", *Agricultural Situation of India*, June 1952.

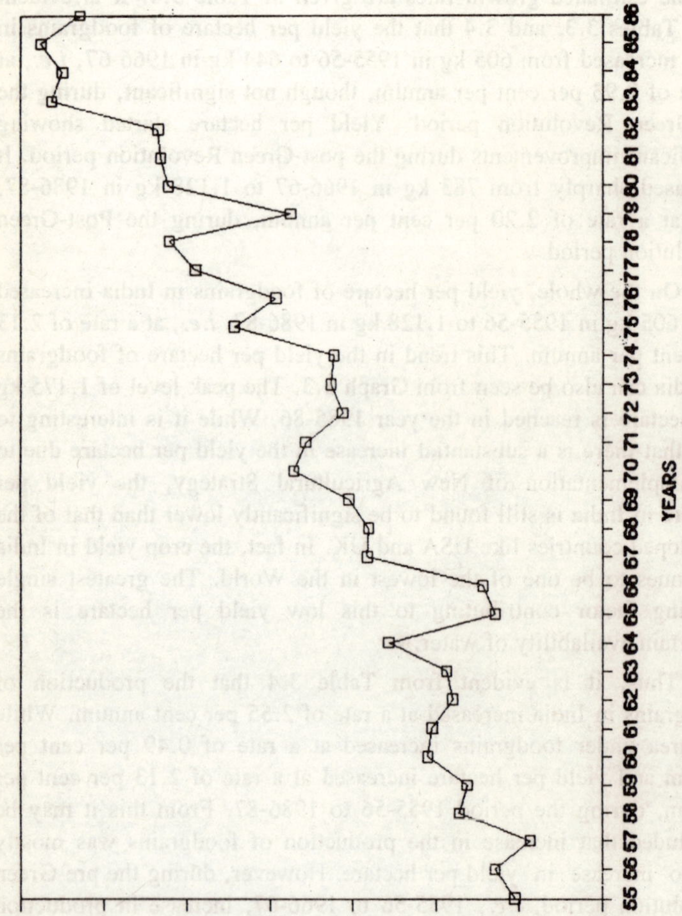

Graph 3.3: Trends in the Yield per Hectare of Area Under Foodgrains in India

hectare (2.20 per cent per annum), as area under foodgrains increased only at a rate of 0.31 per cent per annum.

Wide fluctuations in agricultural production have become a striking element of the Indian economy Production of foodgrains in India showed considerable fluctuations (as can be seen from, Graph 3.2., which shows the trends in the production of foodgrains). The coefficients of variation are found to be 25.12 per cent for production of foodgrains, 4.95 per cent for area and 21.11 per cent for yield during the period 1955-56 to 1986-87. Thus, the instability in the production of foodgrains in India seems to be mostly arising from instability in the yield per hectare of foodgrains (as can be seen from Fig. 3.3). The coefficient of variation is found to be 8.37 per cent for production, 2.55 per cent for area and 7.26 per cent for yield of foodgrains during the pre-Green Revolution period, *i.e.*, 1955-56 to 1966-67. The coefficient of variation is found to be 17.23 per cent for production, 2.57 per cent for area and 14.18 per cent for yield of foodgrains during the post-Green Revolution period, *i.e.*, 1967-68 to 1986-87. Thus, the instability in foodgrains production during the pre-Green Revolution period as well as the post-Green Revolution period is found to be associated with the instability in the yield per hectare. Further, the coefficient of variation is found to be increasing in respect of both production and yield of foodgrains during the period 1955-56 to 1986-87, indicating the increase in the degree of instability. In addition, the coefficient of variation is found to be higher in the post-Green Revolution period compared to the pre-Green Revolution period, in the case of both production of foodgrains as well as yield per hectare of foodgrains, indicating the greater instability in production and yield of foodgrains in the post-Green Revolution period. The study by Shakuntala Mehra confirms the increase in the standard deviation and the coefficient of variation for production of crops during the post-Independence period.[19]

3.4. Cropping Pattern of Foodgrains in India

Foodgrains occupy the first place amongst all the crops in terms of area. The area under foodgrain crops constitutes more than three-

---

19. S.Mehra, *Instability in Indian Agriculture in the Context of the New Technology*, Research Report 25, International Food Policy Research Institute, July 1981. See also P.B.R. Hazell, *Instability in Indian Foodgrain Production*, Research Report 30, International Food Policy Research Institute, May 1982.

Table 3.5 : Cropping Pattern of Foodgrains in India—Area

(Area in Million hectares)

Years	Rice	Wheat	Jowar	Bajra	Maize	Ragi	Small millets	Barley	Total Cereals[a]	Total Pulses[b]	Total area under Food-grains in India
(1)	(2)	(3)	(4)	(5)	(6)	(7)	(8)	(9)	(10)	(11)	(12)
1955-56	31.52	12.37	17.36	11.34	3.69	2.31	5.34	3.42	85.84	23.22	110.56
1956-57	32.28	13.52	16.24	11.25	3.76	2.28	4.98	3.52	87.34	23.32	111.14
1957-58	32.29	11.73	17.31	11.17	4.08	2.42	4.87	3.07	87.82	22.54	109.48
1958-59	33.17	12.62	17.96	11.43	4.27	2.54	5.16	3.31	86.94	24.31	114.76
1959-60	33.82	13.38	17.71	10.69	4.34	2.52	5.15	3.38	90.45	24.83	115.82
1960-61	34.13	12.93	18.41	11.47	4.41	2.52	4.96	3.21	90.99	23.56	115.58
1961-62	34.69	13.57	18.25	11.28	4.51	2.51	4.87	3.31	92.02	24.24	117.23
1962-63	35.69	13.59	18.41	10.96	4.64	2.48	4.77	3.02	99.99	24.27	117.42
1963-64	35.81	13.49	18.38	11.10	4.58	2.47	4.62	2.77	93.58	24.19	117.42
1964-65	35.46	13.42	18.06	11.83	4.62	2.61	4.56	2.68	93.24	23.88	118.11
1965-66	35.47	12.57	17.68	11.97	4.79	2.69	4.56	2.64	94.24	22.72	115.10
1966-67	35.25	12.84	18.05	12.24	5.07	2.32	4.58	2.83	92.39	22.12	115.30
1967-68	36.44	14.99	18.42	12.81	5.58	2.23	4.86	2.38	93.18	22.65	121.42
1968-69	36.97	15.96	18.73	12.05	5.72	2.24	4.75	2.76	96.77	21.26	120.43
1969-70	37.68	16.63	18.61	12.49	5.86	2.78	4.73	2.77	99.17	22.02	123.57

1970-71	37.59	18.24	17.37	12.91	5.85	2.47	4.78	2.56	101.55	22.53	124.32
1971-72	37.76	19.14	16.78	11.77	5.67	2.43	4.48	2.46	100.57	22.15	122.62
1972-73	36.69	19.46	15.51	11.82	5.84	2.33	4.27	2.45	98.36	20.92	119.28
1973-74	38.29	18.58	16.72	13.93	6.02	2.36	4.57	2.65	103.11	23.43	126.54
1974-75	37,89	18.01	16.19	11.29	5.86	2.46	4.47	2.89	99.05	22.02	121.08
1975-76	39.48	20.45	26.09	11.57	6.03	2.63	4.67	2.80	103.73	24.45	128.18
1976-77	38.51	20.92	15.77	10.75	6.00	2.49	4.68	2.24	101.37	22.98	124.36
1977-78	40.28	21.46	16.32	11.10	5.68	2.60	4.57	2.00	104.02	23.49	127.51
1978-79	40.48	22.64	16.15	11.39	5.76	2.71	4.39	1.83	105.35	23.66	129.01
1979-80	39.41	22.17	16.67	10.58	5.72	2.62	4.00	1.77	102.95	22.26	125.21
1980-81	40.15	22.28	15.81	11.66	6.01	2.53	3.98	1.81	104.21	22.46	126.67
1981-82	40.71	22.14	16.60	11.78	5.94	2.61	3.79	1.73	105.30	23.84	129.14
1982-83	38.26	23.57	16.37	10.94	5.72	2.36	3.46	1.49	102.26	22.83	125.09
1983-84	41.24	24.67	16.43	11.83	5.86	2.61	3.61	1.38	107.62	23.54	131.16
1984-85	41.16	23.56	15.44	10.62	5.80	2.39	3.21	1.25	103.93	22.74	126.67
1985-86	41.14	23.00	16.10	16.65	5.80	2.40	3.61	1.37	103.61	24.42	128.02
1986-87	41.77	22.81	15.64	11.22	5.87	2.39	3.09	1.28	104.04	23.09	127.20

Notes: (a) Includes rice, wheat, jowar, bajra, barely and ragi
(b) Includes gram, arhar, mung, masur and urad.

Source: (i) GOI., *Bulletin on Food Statistics*, Delhi, various issues.
(ii) GOI., *Estimates of Area and Production of Principal Crops in India*, Various issues.
(iii) RBI., *Report on Currency and Finance*, various issues, Bombay.

Table 3.6 : Relative Shares in the Area under Foodgrains—Cropwise

(Percentage)

Period	Rice	Wheat	Jowar	Bajra	Maize	Ragi	Small Millets	Barley	Total Cereals	Total Pulses	Total Foodgrains
(1)	(2)	(3)	(4)	(5)	(6)	(7)	(8)	(9)	(10)	(11)	(12)
1955-56	28.5	11.2	15.7	10.3	3.3	2.1	4.8	3.1	77.7	22.3	100.0
1956-57	29.0	12.2	14.6	10.1	3.4	2.1	4.5	3.2	78.6	21.4	100.0
1957-58	29.5	10.7	15.8	10.2	3.7	2.2	4.4	2.8	80.2	19.8	100.0
1958-59	28.9	10.9	15.7	9.9	3.7	2.2	4.5	2.9	75.8	24.2	100.0
1959-60	29.2	11.6	15.3	9.2	3.7	2.1	4.4	2.9	78.1	20.9	100.0
1960-61	29.5	11.2	15.9	9.9	3.8	2.2	4.3	2.8	78.8	21.3	100.0
1961-62	29.6	11.6	15.6	9.6	3.8	2.1	4.2	2.8	78.5	21.5	100.0
1962-63	30.3	11.5	15.6	9.3	3.9	2.1	4.0	2.6	80.9	19.1	100.0
1963-64	30.5	11.5	15.7	9.5	3.9	2.1	3.9	2.4	79.7	20.3	100.0
1964-65	30.9	11.4	15.3	10.0	3.9	2.2	3.9	2.3	78.9	21.1	100.0
1965-66	30.8	10.9	15.4	10.4	3.9	2.3	3.4	2.3	80.4	19.6	100.0
1966-67	30.6	11.1	15.7	10.6	4.2	2.0	3.9	2.5	80.1	19.9	100.0
1967-68	30.0	12.3	15.2	10.6	4.4	1.8	4.0	1.9	81.3	18.7	100.0
1968-69	30.7	13.2	15.6	10.0	4.6	1.9	3.9	2.3	82.3	17.7	100.0

1969-70	30.5	13.5	15.1	10.1	4.7	2.2	3.8	2.2	82.2	17.8	100.0
1970-71	28.8	14.7	14.0	10.4	4.7	2.0	3.8	2.1	81.7	18.3	100.0
1971-72	27.5	15.6	13.7	10.0	4.6	2.0	3.7	2.0	81.9	18.1	100.0
1972-73	30.8	16.3	13.0	10.0	4.9	1.9	3.6	2.1	82.5	17.5	100.0
1973-74	30.3	14.7	13.2	11.0	4.8	1.8	3.6	2.1	81.5	18.5	100.0
1974-75	31.3	14.9	13.4	9.3	4.8	2.0	3.7	2.4	81.8	18.2	100.0
1975-76	30.8	15.9	12.6	9.0	4.7	2.1	3.6	2.2	80.9	19.1	100.0
1976-77	30.9	16.8	12.7	8.6	4.8	2.0	3.8	1.8	81.5	81.5	100.0
1977-78	31.6	16.8	12.8	8.7	4.5	2.0	3.6	1.6	81.6	18.4	100.0
1978-79	31.4	17.5	12.5	8.8	4.5	2.1	3.4	1.4	81.7	18.3	100.0
1979-80	31.5	17.7	13.3	8.4	4.6	2.1	3.2	1.4	82.2	17.8	100.0
1980-81	31.7	17.5	12.5	9.2	4.7	2.0	3.1	1.4	82.3	17.7	100.0
1981-82	31.5	17.1	12.9	9.1	4.6	2.0	3.0	1.3	81.5	18.5	100.0
1982-83	30.6	18.8	13.1	8.7	4.6	2.0	2.8	1.1	81.7	18.3	100.0
1983-84	31.4	18.8	12.5	9.0	4.5	2.0	2.8	1.1	82.1	17.9	100.0
1984-85	32.5	20.0	12.2	8.4	4.6	1.9	2.5	0.9	82.0	18.0	100.0
1985-86	32.1	18.0	12.6	13.0	4.5	1.9	2.5	1.1	80.9	19.1	100.0
1986-87	32.8	17.9	12.3	8.8	4.6	1.9	2.4	1.0	81.8	18.2	100.0

Source : Calculated from data given in Table 3.5.

Table 3.7 : Growth Rates of Area Under Foodgrains—Crop-wise

Period	Rice		Wheat		Jowar		Bajra		Maize	
	Gr.	CV	Gr.	CV	Gr.	CV	Gr.	CV	Gr.	CV
1955-56 to 1966-67	1.22* (7.74)	4.71	0.39 (1.01)	4.56	0.60** (2.37)	3.52	0.60** (2.17)	3.84	2.49* (10.29)	9.17
1967-68 to 1986-87	0.65* (8.44)	4.31	2.17* (9.62)	13.40	-0.65* (3.96)	5.84	-0.79* (3.60)	7.43	0.09 (1.01)	2.21
1955-56 to 1986-87	0.81 (19.00)	7.79	2.52* (17.47)	24.22	-0.44* (5.17)	6.07	-0.07 (0.57)	6.39	1.41* (0.96)	14.21

Period	Ragi		Small Millets		Barley		Total Cereals		Total Pulses		Total area Under Foodgrains	
	Gr.	CV	Gr.	CV	Gr.	CV	Gr.	CV	Gr.	CV	Gr.	CV
1955-56 to 1966-67	0.66 (1.70)	4.97	-1.31 (5.96)	5.37	-2.38* (5.20)	9.90	0.90* (3.70)	4.32	-0.15 (0.49)	3.53	0.52* (3.36)	2.55
1967-68 to 1986-87	0.19 (0.81)	6.08	-2.30* (8.85)	14.19	-4.92* (12.45)	29.54	0.44* (5.21)	3.31	0.37*** (2.69)	4.11	0.31* (4.34)	2.57
1955-56 to 1986-87	0.09 (0.89)	5.61	-1.34* (9.34)	13.43	-3.11* (12.54)	28.24	0.65* (12.66)	6.50	-0.06 (0.77)	4.16	0.49* (12.56)	4.95

Note: Gr. = Exponential growth.
CV = Coefficient of variation.
Figures in brackets represent 't' values.
* Significant at 1 per cent level.
** Significant at 5 per cent level.
*** Significant at 10 per cent level.

fourths of the total cropped area.[20] It is indicative of subsistence farming in India and of the scope for commercialisation. The outstanding feature of cropping pattern in India is the wide variety of crops and the preponderence of food over non-food crops. The major food crops include: rice, wheat, jowar, bajra, maize, ragi, barely, small millets and pulses, which may further be grouped into (*a*) cereals, (*b*) coarse cereals and (*c*) pulses. The cropping pattern of foodgrains in India is dominated by cereals, which account for about 80 per cent of total food production.[21]

An attempt is made in this Section to analyse the trends in the cropping pattern of foodgrains in India in order to examine the changes that have taken place in the cropping pattern. Table 3.5 shows the trends in the cropping pattern of foodgrains in India. The relative share of each of these crops in the total area under foodgrains is shown in Table 3.6. Exponential growth rates are estimated for the area under each category of crop. Coefficient of variation has been calculated for each crop to analyse the fluctuations in area. The coefficients of variation and the estimated growth rates are shown in Table 3.7.

It is evident from Tables 3.5 and 3.7 that area under rice increased from 31.52 million hectares in 1955-56 to 41.77 million hectares in 1986-87, *i.e.*, at a rate of 0.81 per cent per annum. Growth rate of area under rice showed a decline from 1.22 per cent per annum during the pre-Green Revolution period to 0.65 per cent per annum during the post-Green Revolution period.

Area under wheat has been increasing since the inception of planning. Green Revolution contributed greatly to the richness of the crop and also to the extension of area under this crop. With the introduction of new technological innovations, dry areas are also being brought under wheat cultivation. Area under wheat increased from 12.37 million hectares in 1955-56 to 22.81 million hectares in 1986-87, *i.e.*, at a rate of 2.52 per cent annum. Area under wheat remained more or less stagnant during the pre-Green Revolution period but increased at a significant rate of 2.27 per cent per annum during the post-Green Revolution period.

---

20. S.C. Chandra, *Agricultural Price Policy in India*, Chugh Pub., Allahabad, 1985, p.16.
21. RBI., *Report on Currency and Finance 1986-87, Vol.* I, Economic Review, Bombay, 1986, p. 18.

Jowar and bajra are two of the important millet crops which are grown in India on areas which have relatively less rainfall and irrigational facilities. Area under jowar declined from 17.36 million hectares in 1955-56 to 15.64 million hectares in 1986-87, *i.e.*, at a rate of 0.44 per cent per annum Area under bajra declined from 11.34 million hectares to 11.22 million hectares *i.e.*, at a rate of 0.07 per cent per annum, though not significant, during the same period.

Area under maize increased from 3.69 million hectares in 1955-56 to 5.87, *i.e.*, at a rate of 1.41 per cent per annum. Growth rate is found to be significant during the pre-Green Revolution period, but is not found to be significant during the post-Green Revolution period. Area under ragi remained more or less stagnant in the pre-Green Revolution period, as well as in the post-Green Revolution period. On the whole, area under ragi increased only from 2.31 million hectares in 1955-56 to 2.39 million hectares in 1986-87, *i.e.*, at a rate of 0.09 per cent per annum, but not significant. Area under small millets declined from 5.34 million hectares in 1955-56 to 3.09 million hectares in 1986-87, *i.e.*, at a rate of 1.34 per cent per annum, while area under barley declined from 3.42 million hectares in 1955-56 to 1.28 million hectares in 1986-87, *i.e.*, at a rate of 3.11 per cent per annum.

Area under total cereals remained relatively stagnant during the pre-Green Revolution period. On the whole, it increased at a rate of 0.65 per cent per annum during the period 1955-56 to 1986-87. Area under total pulses showed a decline during the pre-Green Revolution period, while it showed only a marginal increase during the post-Green Revolution period.

Total area under foodgrains in India (area under total cereals plus area under total pulses) increased at a rate of 0.52 per cent per annum, during the pre-Green Revolution period and at a rate of 0.31 per cent per annum, during the post-Green Revolution period. On the whole, area under total foodgrains increased from 110.56 million hectares in 1955-56 to 127.20 million hectares in 1986-87, *i.e.*, at a rate of 0.49 per cent annum.

It is evident that during the period 1955-56 to 1986-87, changes have taken place to some extent in the cropping pattern of foodgrains in India, with area under rice, wheat and maize showing an increase and area under jowar, small millets and barely showing a decline.

Total area under foodgrains, however, has not shown much increase, the growth rate being less than 0.50 per cent per annum.

Table 3.6 indicates that the area under rice accounted for 28.5 per cent of the total area under foodgrains in 1955-56 and 32.8 per cent in 1986-87, while the relative share of area under wheat increased from 11.2 per cent in 1955-56 to 17.9 per cent in 1986-87. These two crops together accounted for 39.7 per cent in 1955-56 and 50.7 per cent in 1986-87, of the total area under foodgrains.

Area under total cereals accounted for 77.7 per cent of the total area under foodgrains in 1955-56 and 81.8 per cent in 1986-87, while the share of area under total pulses in total foodgrains showed a decline from 22.3 per cent in 1955-56 to 18.2 per cent in 1986-87, indicating the changes in the cropping pattern of foodgrains in India.

Coefficient of variation is found to be highest in the case of barely (28.24 per cent) followed by wheat (24.22 per cent), maize (14.21 per cent), small millets (13.43 per cent), rice (7.79 per cent), bajra (6.39 per cent), Jowar (6.07 per cent) and ragi (5.61 per cent) during the period 1955-56 to 1986-87. Coefficient of variation is found to be 6.50 per cent for total cereals, 4.16 per cent for total pulses and 4.95 per cent for total foodgrains during the period 1955-56 to 1986-87.

The official agricultural pricing policy has made rice and wheat virtually the cash crops by increasing their support price every year besides providing effective marketing support. The newly irrigated areas are also increasingly being used for the cultivation of cereals, pushing pulses and oilseeds to more and more marginal lands. The area under the small millets and other major crops has actually declined showing a shift from minor to major crops.

3.5. Pattern of Production of Foodgrains in India

Trends in the pattern of production of foodgrains in India are shown in Table 3.8 . The coefficients of variation for each of the individual crops and the estimated exponential growth rates for each of the individual crops are shown in Table 3.9. The relative shares of each of the individual crops in the total production of foodgrains are shown in Table 3.10.

Table 3.8 : Pattern of Production of Foodgrains in India

(Million Tonnes)

Period	Rice	Wheat	Jowar	Bajra	Maize	Ragi	Small Millets	Barley	Total Cereals	Total Pulses	Total Foodgrains
(1)	(2)	(3)	(4)	(5)	(6)	(7)	(8)	(9)	(10)	(11)	(12)
1955-56	26.8	8.9	6.9	3.8	2.8	1.9	2.7	2.8	56.6	12.7	69.3
1956-57	29.7	9.5	7.8	2.8	3.6	1.8	1.9	2.8	59.9	12.5	72.4
1957-58	25.9	7.9	8.6	3.9	3.7	1.8	1.7	2.3	56.0	10.6	66.6
1958-59	30.9	9.9	9.7	3.9	3.5	1.9	2.9	2.8	65.1	13.7	78.8
1959-60	31.7	10.4	8.6	3.5	4.5	1.9	2.0	2.7	65.3	11.8	77.1
1960-61	34.9	10.9	9.8	3.3	4.1	1.8	1.9	2.8	69.5	12.8	82.3
1961-62	33.9	12.5	8.9	3.8	4.3	2.0	2.1	3.4	71.0	11.7	82.7
1962-63	32.3	10.8	9.8	3.9	4.6	2.0	1.9	3.4	68.6	11.6	80.2
1963-64	36.9	9.9	9.2	3.9	4.4	2.0	2.0	2.0	70.5	10.1	80.6
1964-65	39.3	12.3	9.7	4.5	4.7	2.0	1.9	2.6	77.0	12.4	89.4
1965-66	30.6	10.4	7.6	3.8	4.8	1.3	1.6	2.4	62.4	9.9	72.3
1966-67	30.4	11.4	9.2	4.5	4.9	1.6	1.5	2.4	65.9	8.3	74.2
1967-68	37.6	16.5	10.1	5.4	6.4	1.7	1.8	3.5	83.0	11.1	95.1
1968-69	39.5	18.7	9.8	3.8	5.7	1.6	1.8	2.4	83.3	10.7	94.0

1969-70	40.4	20.1	9.7	5.3	5.7	2.1	1.7	2.7	87.7	11.8	99.5
1970-71	42.2	29.8	8.1	8.0	7.5	2.2	1.9	2.8	96.5	11.9	108.4
1971-72	43.1	26.4	7.7	5.3	5.1	2.2	1.7	2.6	94.1	11.0	105.1
1972-73	39.3	24.7	6.9	3.9	6.4	1.9	1.6	2.4	87.1	9.9	97.0
1973-74	44.1	21.8	9.1	7.5	5.8	2.2	1.9	2.4	94.7	10.0	104.7
1974-75	39.6	24.1	10.4	3.3	5.6	2.1	1.6	3.1	89.8	10.0	99.8
1975-76	48.7	28.9	9.5	5.7	7.3	2.8	1.9	3.2	108.0	13.0	121.0
1976-77	41.9	29.0	10.5	5.9	6.4	2.1	1.8	2.3	99.9	11.3	111.2
1977-78	52.6	31.8	12.1	4.7	5.9	2.9	2.1	2.3	114.4	12.0	126.4
1978-79	53.8	35.6	11.4	5.6	6.2	3.2	1.9	2.1	119.8	12.1	131.9
1979-80	42.4	31.8	11.7	3.9	5.6	2.7	1.4	1.6	101.1	8.6	109.7
1980-81	53.6	36.3	10.4	5.3	6.9	2.4	1.6	2.3	118.8	10.8	129.6
1981-82	53.3	37.5	12.1	5.5	6.9	2.9	1.7	2.0	121.9	11.4	133.3
1982-83	47.1	42.8	10.8	5.1	6.6	2.2	1.2	1.8	117.6	11.9	129.5
1983-84	60.9	45.9	12.9	5.7	7.9	2.8	1.8	1.8	139.5	12.9	152.4
1984-85	58.3	44.4	11.8	6.6	6.4	2.8	1.8	1.6	133.7	11.8	145.5
1985-86	63.8	47.1	10.2	3.7	6.6	2.5	1.2	1.9	137.0	13.4	150.4
1986-87	60.6	44.3	9.2	4.5	7.5	2.7	1.1	1.7	131.7	11.7	143.4

Note : Production in terms of cleaned rice. production relates to October-September.

Source: (1) GOI., Ministry of Agriculture, *Estimates of Area and Production of Principal Crops in India*, various issues.
(2) GOI., *Bulletin on Food Statistics*, various issues.
(3) GOI., *Economic Survey*, various issues.
(4) RBI., *Report on Currency and Finance*, Vol, II, Statistical Statement, various issues.

Table 3.9 : Growth Rates of Production of Foodgrains_Crop-wise

Period	Rice		Wheat		Jowar		Bajra		Maize		Ragi	
	Gr.	CV	Gr.	CV	Gr.	CV	Gr.	CV	Gr.	CV.	Gr.	CV
1955-56 to 1966-67	1.92** (2.21)	12.12	2.41* (2.83)	12.73	1.33 (1.47)	10.86	2.12** (2.37)	12.19	4.08* (6.11)	15.29	-1.25 (1.22)	11.23
1967-68 to 1986-87	2.51* (7.76)	17.33	5.24* (16.80)	30.80	1.42* (2.62)	15.22	-0.32 (0.34)	23.32	0.91** (2.21)	11.66	2.27* (4.34)	18.19
1955-56 to 1986-87	2.49* (14.58)	25.10	6.11* (23.98)	54.69	1.04 (4.27)	15.59	1.36* (3.31)	25.97	2.37* (9.82)	23.57	1.63* (5.91)	20.92

Period	Small Millets		Barley		Total Cereals		Total Pulses		Total Foodgrains	
	Gr.	CV	Gr.	CV	Gr.	CV	Gr.	CV	Gr.	CV
1955-56 to 1966-67	-3.20** (2.41)	20.42	-1.06 (0.82)	15.23	1.69* (2.63)	9.45	-2.57* (2.77)	13.23	1.06 (1.62)	8.37
1967-68 to 1986-87	-1.59* (2.72)	15.84	-3.15* (5.93)	23.22	2.74* (11.47)	17.35	0.57 (1.44)	10.23	2.51* (10.14)	16.23
1955-56 to 1986-87	-1.34* (4.49)	20.0'	-1.54* (4.93)	21.18	2.90* (21.15)	27.94	-0.03 (0.15)	11.27	2.55* (17.15)	25.12

Note: Gr. = Exponential growth rate.
CV = Coefficient of variation.
Figures in brackets represent 't' values.
* Significant at 1 per cent level.
** Significant at 5 per cent level.
*** Significant at 10 per cent level.

It is evident from Tables 3.8 and 3.9 that production of rice increased from 26.8 million tonnes in 1955-56 to 60.6 million tonnes in 1986-87, *i.e.*, at rate of 2.49 per cent per annum. Production of wheat increased from 8.9 million tonnes in 1955-56 to 44.3 million tonnes in 1986-87, *i.e.*, at a rate of 6.11 per cent per annum. Production of jowar increased from 6.9 million tonnes in 1955-56 to 9.2 million tonnes in 1986-87, *i.e.*, at a rate of 1.04 per cent per annum. Production of bajra increased from 3.8 million tonnes in 1955-56 to 4.5 million tonnes 1986-87, *i.e.*, at a rate of 1.36 per cent per annum. Production of maize increased from 2.8 million tonnes in 1955-56 to 7.5 million tonnes in 1986-87, *i.e.*, at a rate of 2.37 per cent per annum. Production of ragi increased from 1.9 million tonnes in 1955-56 to 2.7 million tonnes in 1986-87, *i.e.*, at a rate of 1.63 per cent per annum. On the other hand, production of small millets declined from 2.7 million tonnes in 1955-56 to 1.1 million tonnes in 1986-87, *i.e.*, at a rate of 1.34 per cent per annum. Production of barley declined from 2.8 million tonnes in 1955-56 to 1.7 million tonnes in 1986-87, *i.e.*, at a rate of 1.54 per cent per annum.

It is evident from Table 3.9, that there are variations in the growth rates of the different foodgrain crops. During the pre-Green Revolution period 1955-56 to 1966-67, rice, wheat, bajra and maize showed positive growth rates while small millets showed a decline. The highest growth rate of 4.08 per cent per annum is exhibited by maize followed by wheat with 2.41 per cent per annum, bajra with 2.12 per cent per annum and rice with 1.92 per cent per annum.

During the post-Green Revolution period 1967-68 to 1986-87, rice, wheat, jowar, maize and ragi showed positive and significant growth rates while small millets and barley showed negative growth rates. The highest growth rate of 5.24 per cent per annum is exhibited by wheat followed by 2.51 per cent per annum of rice, 2.27 per cent per annum of ragi, 1.42 per cent per annum of jowar and 0.91 per cent per annum of maize.

On the whole, during the period 1955-56 to 1986-87, rice, wheat, jowar, bajra, maize and ragi showed positive and significant growth rates while small millets and barley showed negative and significant growth rates. Wheat showed the highest growth rate of 6.11 per cent per annum followed by rice with 2.49 per cent, maize with 2.37 per cent, ragi with 1.63 per cent, bajra with 1.36 per cent

Table 3.10 : Relative Share of Production of Foodgrains : Crop-wise

(Percentage)

Period (1)	Rice (2)	Wheat (3)	Jowar (4)	Bajra (5)	Maize (6)	Ragi (7)	Barley (8)	Small Millets (9)	Total Cereals (10)	Total Pulses (11)	Total Foodgrains (12)
1955-56	38.67	12.84	9.96	5.48	4.04	2.74	4.04	3.90	81.67	18.83	100.00
1956-57	41.02	13.12	10.77	3.87	4.97	2.49	3.87	2.62	82.73	17.27	100.00
1957-58	38.89	11.86	12.91	5.86	5.56	2.70	3.45	2.55	84.08	15.92	100.00
1958-59	39.21	12.56	12.31	4.95	4.44	2.41	3.55	3.68	82.61	17.39	100.00
1959-60	41.12	13.49	11.15	4.54	5.84	2.46	3.50	2.59	84.70	15.30	100.00
1960-61	42.41	13.24	11.91	4.01	4.98	2.19	3.40	2.31	84.45	15.55	100.00
1961-62	40.99	15.11	10.76	4.59	5.20	2.42	4.11	2.54	85.85	14.15	100.00
1962-63	40.15	13.47	12.22	4.86	5.74	2.49	4.24	2.37	85.54	14.46	100.00
1963-64	45.78	12.28	11.41	4.84	5.46	2.48	2.48	2.48	87.47	12.53	100.00
1964-65	45.96	13.76	10.85	5.03	5.26	2.24	2.91	2.13	86.13	13.87	100.00
1965-66	42.32	14.38	10.51	5.26	6.64	1.80	3.32	2.21	86.31	13.69	100.00
1966-67	40.97	15.36	12.40	6.06	6.60	2.16	3.23	2.02	88.81	11.19	100.00
1967-68	39.54	17.35	10.62	5.68	6.73	1.79	3.68	1.89	87.28	12.72	100.00
1968-69	42.02	19.89	10.43	4.04	6.06	1.70	2.55	1.91	88.62	11.38	100.00
1969-70	40.60	20.20	9.75	5.33	5.73	2.11	2.71	1.71	88.14	11.86	100.00

1970-71	38.93	21.96	7.47	7.38	6.92	2.03	2.58	1.75	89.02	10.98	100.00
1971-72	41.01	25.12	7.33	5.04	4.85	2.09	2.47	1.62	89.53	10.47	100.00
1972-73	40.52	25.46	7.11	4.02	6.60	1.96	2.47	1.65	89.79	10.21	100.00
1973-74	42.12	20.82	8.69	7.16	5.54	2.10	2.29	1.81	90.45	9.55	100.00
1974-75	39.68	24.15	10.42	3.31	5.61	2.10	3.11	1.60	89.98	10.02	100.00
1975-76	40.25	23.88	7.85	4.71	6.03	2.31	2.64	1.57	89.26	10.74	100.00
1976-77	37.68	26.08	9.44	5.31	5.76	1.89	2.07	1.62	89.84	10.16	100.00
1977-78	41.61	25.16	9.57	3.72	4.67	2.29	1.82	1.66	90.51	9.49	100.00
1978-79	40.79	26.99	8.64	4.25	4.70	2.43	1.59	1.44	90.83	9.17	100.00
1979-80	38.65	28.99	10.67	3.56	5.10	2.46	1.46	1.28	92.16	7.84	100.00
1980-81	41.36	28.01	8.02	4.09	5.32	1.85	1.77	1.23	91.67	8.33	100.00
1981-82	39.98	28.13	9.08	4.13	5.18	2.18	1.50	1.28	91.45	8.55	100.00
1982-83	36.37	33.05	8.34	3.94	5.10	1.70	1.39	0.93	90.81	9.19	100.00
1983-84	39.96	30.12	8.46	3.74	5.18	1.84	1.18	1.18	91.54	4.46	100.00
1984-85	40.07	30.52	8.11	5.54	4.40	1.92	1.10	1.24	91.89	8.11	100.00
1985-86	42.42	31.32	6.78	2.46	4.39	1.66	1.26	0.80	91.09	8.91	100.00
1986-87	42.26	30.89	6.42	3.14	5.23	1.88	1.19	0.77	91.84	8.16	100.00

Source : Calculated from Table 3.8.

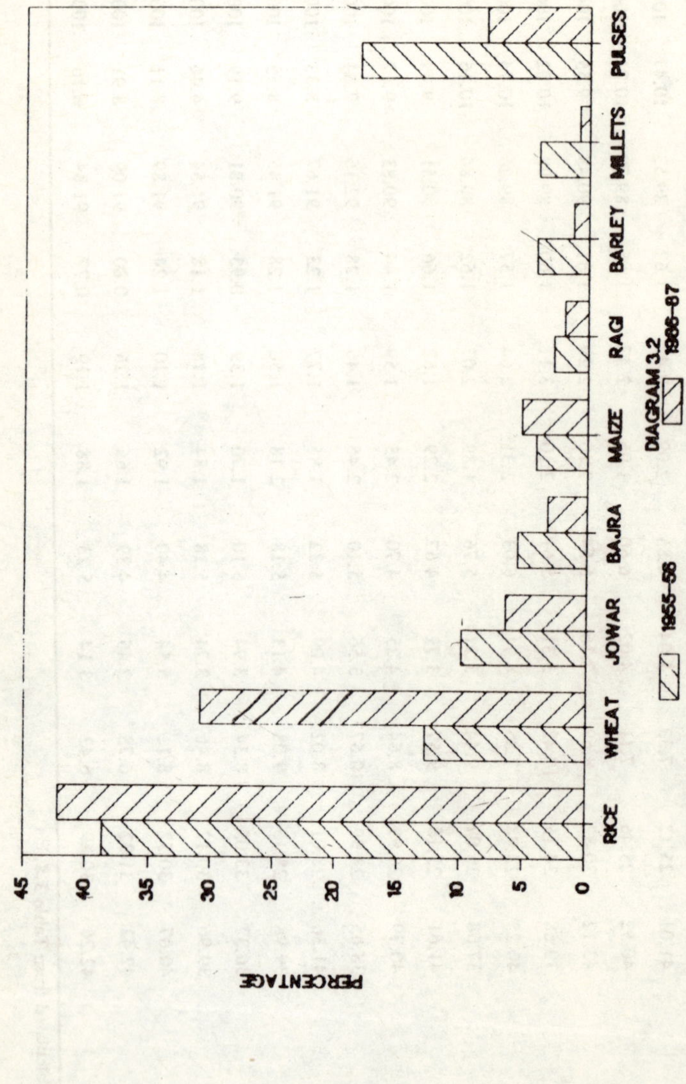

Diagram 3.2: Crop-wise Relative Shares in the Production of Foodgrains in India

and jowar with 1.04 per cent per annum. Among these, the growth rate of wheat alone (6.11 per cent per annum) is found to be higher than the growth rate of 2.55 per cent per annum of total foodgrains in India.

Table 3.9 also gives the coefficients of variation of the individual foodgrain crops. It is evident from the Table that during the period 1955-56 to 1986-87, the highest degree of instability is found in the case of wheat with a coefficient of variation of 54.69 per cent, followed by bajra with 25.97 per cent, rice with 25.10 per cent, maize with 23.57 per cent, barely with 21.18 per cent, ragi with 20.92 per cent, small millets with 20.01 per cent and jowar with 15.59 per cent. The coefficient of variation of total cereals is found to be 27.94 per cent, which is higher than 11.27 per cent of total pulses. The coefficient of variation is found to be 25.12 per cent for the production of total foodgrains in India.

The coefficient of variation is found to be greater in the post-Green Revolution period 1967-68 to 1986-87 compared to the pre-Green Revolution period 1955-56 to 1966-67, except in the case of maize, small millets and total pulses. Thus, there is a great degree of instability in the production of total foodgrains in India. There is also an increase in the instability of the production of foodgrains in the Post-Green Revolution period, 1967-68 to 1986-87. However, there are variations in the degree of instability in the production of the different foodgrain crops, in the different periods.

It is evident from Table 3.10, that rice is the major foodgrain produced in India, accounting for 42.26 per cent, followed by wheat with 30.89 per cent of the total foodgrains produced in India in 1986-87. Together, rice and wheat accounted for 73.15 per cent of the total foodgrains production in 1986-87. There are some changes in the pattern of production of foodgrains in India during the period 1955-56 to 1986-87. There is a rise in relative share of rice, wheat and maize, while there is a decline in the relative share of jowar, bajra, ragi, barley and small millets. Thus, there is a shift towards the production of superior grains from coarse-grains. There is also a shift towards greater production of cereals compared to pulses. This would have been the result of relatively lower economic returns for pulses production.[22]

---

22. S.N. Kulshreshtha and D.D. Tiwari, "Agriculture in India—Problems and Prospects", in J.S. Uppal (ed.) *India's Economic Problems: An Analytical Approach*, Tata McGraw-Hill Publishing Company Ltd., New Delhi, p.101.

The output-mix of cereal crops has undergone important changes over the years because of differential rates of growth recorded by component crops. Rice still remains the predominant cereal, though its position has been very much eroded by the emergence of wheat as a major crop. Similarly maize has recorded gains while jowar has recorded substantial losses.

Thus, changes have taken place in the pattern of production of foodgrains in India during the period under study. This is considered to be one of the important characterstic features of agricultural production in India. The relative shares of jowar, bajra, ragi, barley, millets and pulses showed a decline, while the relative share of rice, wheat and maize showed an increase, (as can be seen from Diagram 3.2). The growth of output of foodgrains has been selective in respect of particular foodgrains and in particular periods.

3.6. Trends in the Foodgrains Prices in India

In countries categorised as 'high food-drain economies',[23] foodgrain prices are a crucial factor shaping the growth of foodgrain production, intersectoral distribution of income, inter-class distribution of income, intersectoral resource transfers and overall growth of the economy. Trends in the wholesale price indices of foodgrains in India are shown in Table 3.11. The estimated exponential growth rates are shown in Table 3.12, which also gives the coefficients of variation.

It is evident from Tables 3.11 and 3.12 that the index number of wholesale prices of creals (with 1970-71 = 100) increased from 36.9 in 1955-56 to 276.4 in 1986-87, *i.e.*, at a rate of 6.33 per cent per annum. During this period, index number of wholesale prices of pulses increased substantially from 28.1 (with base 1970-71 = 100) to 407.2 in 1986-87, *i.e.*, at a rate of 8.47 per cent per annum. Index number of wholesale prices of total foodgrains in India increased from 35.20 in 1955-56 (with 1970-71 = 100) to 298.60 in 1986-87, *i.e.*, at a rate of 6.79 per cent per annum.

---

23. 'High food-drain economies', are those in which expenditure on food, especially foodgrains, forms the bulk of total consumption budget. See T.W. Schultz, *The Economic Organisation of Agriculture*, and M.S. Mudahar, "Foodgrain Prices and Economic Growth", *Indian Journal of Agricultural Economics* Vol. 31, No. 2, April-June 1976, pp. 16-30.

Table 3.11 : Index Number of Wholesale Prices of Foodgrains in India

(Base 1970-71 = 100)

Year	Cereals +	Pulses*	Total[a] Foodgrains
(1)	(2)	(3)	(4)
1955-56	36.9	28.1	35.20
1956-57	47.1	36.6	45.02
1957-58	49.4	37.3	47.00
1958-59	50.9	47.1	51.19
1959-60	48.9	42.3	49.22
1960-61	51.1	42.1	49.26
1961-62	50.1	41.6	48.54
1962-63	51.5	48.6	50.96
1963-64	56.1	54.0	55.70
1964-65	67.7	80.1	70.35
1965-66	103.4	80.0	74.61
1966-67	87.1	93.5	88.44
1967-68	103.4	136.5	110.44
1968-69	97.8	92.7	97.19
1969-70	100.9	99.6	100.67
1970-71	100.0	100.0	100.00
1971-72	101.9	110.7	103.40
1972-73	115.8	137.9	119.50
1973-74	134.8	176.9	141.90
1974-75	191.8	215.7	195.80
1975-76	172.6	181.6	174.10
1976-77	154.1	145.7	152.70
1977-78	161.3	215.2	170.40
1978-79	157.6	247.1	172.60
1979-80	173.5	244.2	185.40
1980-81	195.1	338.7	216.70
1981-82	216.9	338.7	237.40
1982-83	238.0	292.4	263.90
1983-84	259.1	374.4	266.50
1984-85	245.0	430.4	276.20
1985-86	261.7	463.4	295.80
1986-87	276.4	407.2	298.60

Notes: + includes rice, wheat, jowar, bajra, maize and ragi.
* includes gram, arhar, mung, masur and urad.
a arrived by combining Cereals and Pulses sub-groups by weighted average.

Source : (1) GOI., *Bulletin on Food Statistics*, Delhi, various issues.
(2) RBI., *Report on Currency and Finance*, Bombay, various issues.

Table 3.12 : Growth Rates of Wholesale Prices of Foodgrains in India

Period	Cereals		Pulses		Foodgrains	
	Gr.	CV	Gr.	CV	Gr.	CV
1955-56 to 1966-67	5.80* (4.39)	30.43	6.84* (4.63)	33.65	5.08* (5.16)	24.05
1967-68 to 1986-87	5.86* (14.12)	35.03	8.41* (12.99)	49.77	6.39* (15.28)	37.86
1955-56 to 1986-87	6.33* (26.08)	56.86	8.47* (26.19)	76.45	6.79* (30.18)	61.24

Note: Gr. = Exponential growth rate.
CV = Coefficient of Variation.
* Significant at 1 per cent level.
Figures in brackets represent 't' values.

There is a sharp rise in the prices of foodgrains due to various factors like: considerable variations in foodgrains production due to drought and flood conditions, growing population, urbanisation, changes in incomes, speculative hoardings by traders and rich farmers, effect of fiscal policies and other factors like increase in the cost of agricultural inputs, marketing and warehousing facilities.

Prices of cereals, pulses and foodgrains have not only shown an increasing trend but also considerable fluctuations. The coefficient of variation is found to be 56.86 per cent of cereals, 76.45 per cent for pulses and 61.24 per cent for total foodgrains during the period 1955-56 to 1986-87.

Prices of pulses have shown greater fluctuations than those of cereals. Fluctuations in prices of cereals, pulses and foodgrains are found to be higher in the post-Green Revolution period (35.03 per cent, 49.77 per cent and 37.86 per cent in terms of coefficient of variation, respectively), compared to the pre-Green Revolution period, (30.43 per cent, 33.65 per cent and 24.05 per cent in terms of coefficient of variation, respectively).

Prices have shown a continued tendency to rise. Although foodgrains production has increased, prices of foodgrains have not fallen steeply because of the price support extended by Government through its procurement policy.[24]

24. GOI., *Economic Survey 1970-71*, p.1.

3.7. Trends in Per Capita Net Availability of Foodgrains in India

Trends in the per capita per day net availability of foodgrains in India are shown in Table 3.13. The estimated exponential growth rates and the coefficients of variation are shown in Table 3.14.

It is evident from Tables 3.13 and 3.14 that per capita net availability of cereals increased from 360.5 grams per day in 1955-56 to 436.2 grams per day in 1986-87, *i.e.*, at a rate of 0.51 per cent per annum. The per capita net availability of pulses declined from 70.4 grams per day in 1955-56 to 35.9 grams per day in 1986-87, *i.e.*, at a rate of 2.01 per cent per annum. In a vegetarian country like India, pulses are a rich source of protein. However, recent statistics have shown that, instead of increasing, the per capita pulses intake in the country is decreasing during the period under study.

Per capita net availability of foodgrains remained more or less stagnant in the pre-Green Revolution period as well as in the Post-Green Revolution period, with growth rates found to be not significant. On the whole, per capita net availability of foodgrains in India increased from 430.9 grams per day in 1955-56 to 472.1 grams per day in 1986-87, *i.e.*, at a rate of 0.24 per cent per annum, with a peak level of 480.2 grams per day in 1964-65 and with considerable fluctuations, as can be seen from Graph 3.4. The level of per capita net availability of foodgrains per day is lower than the minimum nutrition requirements of 594 grams per day, as specified by the Indian Council of Medical Research.[25]

3.8. Trends in the Procurement, Public Distribution and Stocks of Foodgrains in India

Government's buying and selling operations are called as bufferstock operations.[26] It is a stock of foodgrains fed from the domestically procured food and from imports. Trends in the procurement of foodgrains are shown in Col.(2) of Table 3.15. Estimated exponential growth rates and coefficients of variation are shown in Table 3.16. It is evident from Tables 3.15 and 3.16 that foodgrains procured by the Government increased from 0.04 million

---

25. ICMR (Indian Council of Medical Research), *Recommended Balanced Diet*, 1966.
26. A.N. Agrawal, *Indian Economy*, Allied Publishers, New Delhi, 1975, p.374.

Table 3.13 : Per Capital Per Day Net Availability of Foodgrains in India

(in Grams)

Year (1)	Cereals (2)	Pulses (3)	Foodgrains (4)
1955-56	360.5	70.4	430.9
1956-57	375.1	72.0	447.1
1957-58	350.2	58.7	408.9
1958-59	393.5	75.0	468.5
1959-60	382.8	65.5	448.9
1960-61	399.7	69.0	468.7
1961-62	399.0	62.0	461.0
1962-63	384.0	59.8	433.8
1963-64	401.0	51.0	452.0
1964-65	418.6	61.6	480.2
1965-66	360.0	48.2	408.2
1966-67	361.7	39.7	401.4
1967-68	404.1	56.0	460.1
1968-69	397.9	47.3	445.2
1969-70	403.1	51.9	455.0
1970-71	417.6	51.2	460.1
1971-72	419.1	47.0	467.3
1972-73	380.5	41.1	423.7
1973-74	410.4	40.8	451.2
1974-75	365.8	39.7	408.6
1975-76	373.8	50.5	457.8
1976-77	386.4	43.3	429.6
1977-78	422.5	45.5	468.0
1978-79	431.8	44.7	476.5
1979-80	379.5	30.9	410.4
1980-81	416.2	37.5	453.7
1981-82	414.8	39.2	455.0
1982-83	396.9	39.5	436.4
1983-84	436.1	41.8	477.9
1984-85	415.6	38.1	453.7
1985-86	434.2	41.9	476.1
1986-87	436.2	35.9	472.1

Source : GOI., *Economic Survey*, Delhi, various issues.

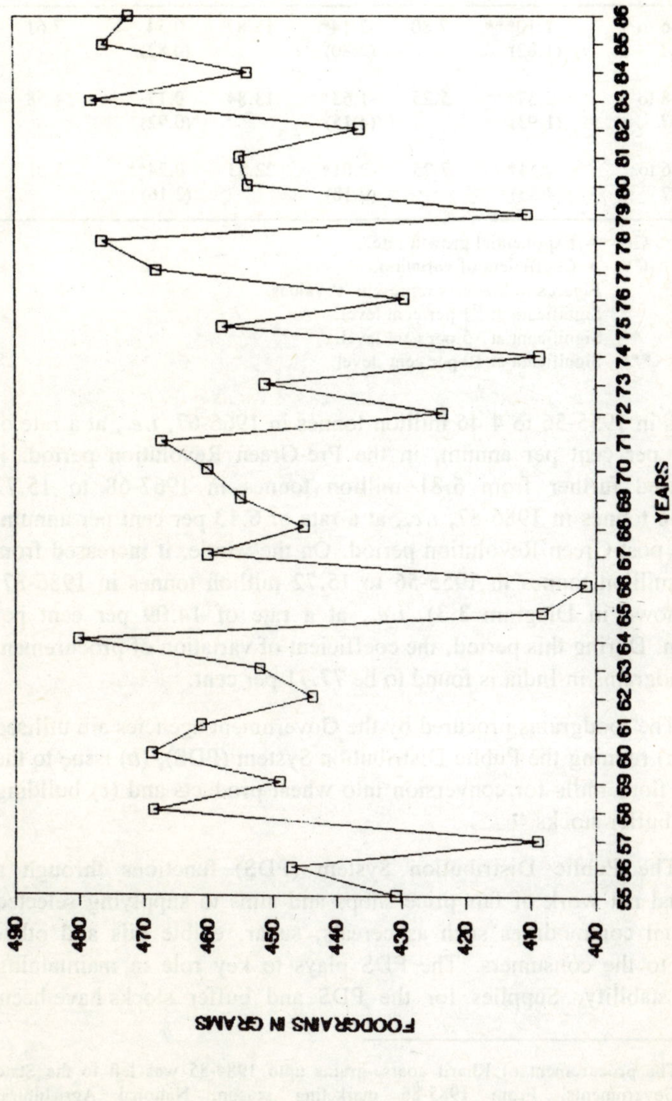

Graph 3.4: Trends in the Per Capita Net Availability of Foodgrains in India

Table 3.14 : Growth Rates of Per Capita Per Day Net Availability of Foodgrains

Period	Total Cereals		Total Pulses		Total Foodgrains	
	Gr.	CV	Gr.	CV	Gr.	CV
1955-56 to 1966-67	1.10*** (1.82)	7.80	-3.14* (2.80)	15.87	0.54 (0.83)	7.61
1967-68 to 1986-87	0.37*** (1.93)	5.25	-1.63* (4.15)	13.84	0.17 (0.92)	4.58
1955-56 to 1986-87	0.51* (4.53)	7.25	-2.01* (9.18)	22.53	0.24** (2.16)	6.01

Note: Gr. = Exponential growth rate.
CV = Coefficient of variation.
Figures in brackets represent 't' values.
* Significant at 1 per cent level.
** Significant at 5 per cent level.
*** Significant at 10 per cent level.

tonnes in 1955-56 to 4.46 million tonnes in 1966-67, *i.e.*, at a rate of 30.24 per cent per annum, in the Pre-Green Revolution period. it increased further from 6.81 million tonnes in 1967-68 to 15.72 million tonnes in 1986-87, *i.e.*, at a rate of 6.13 per cent per annum, in the post-Green Revolution period. On the whole, it increased from 0.44 million tonnes in 1955-56 to 15.72 million tonnes in 1986-87, (as shown in Diagram 3.3), *i.e.*, at a rate of 14.09 per cent per annum. During this period, the coefficient of variation of procurement of foodgrains in India is found to be 77.71 per cent.

The foodgrains procured by the Government agencies are utilised for (*a*) running the Public Distribution System (PDS); (*b*) issue to the roller flour mills for conversion into wheat products and (*c*) building up of buffer stocks.[27]

The Public Distribution System (PDS) functions through a national net work of fair price shops and aims to supplying selected essential commodities such as cereals, sugar, edible oils and other items to the consumers. The PDS plays to key role in maintaining price stability. Supplies for the PDS and buffer stocks have been

27. The procurement of Kharif coarse-grains upto 1984-85 was left to the State Governments. From 1985-86 marketing season, National Agricultural Cooperative Marketing Federation (NCAMF) has been designated as the model agency for procurement of coarse-grains.

Table 3.15: Trends in the Procurement, Public Distribution and Stocks of Foodgrains in India

(Million Tonnes)

Year (1)	Total Procurement (2)	Total Public Distribution (3)	Total* Stocks (4)
1955-56	0.04	2.08	0.32
1956-57	0.30	3.05	1.18
1957-58	0.53	3.98	0.91
1958-59	1.81	5.16	1.39
1959-60	1.28	4.94	2.80
1960-61	0.54	3.98	2.64
1961-62	0.48	4.37	2.28
1962-63	0.75	5.18	2.26
1963-64	1.43	8.67	1.02
1964-65	4.03	10.08	2.08
1965-66	4.01	14.09	2.22
1966-67	4.46	13.17	1.96
1967-68	6.81	10.22	3.99
1968-69	6.38	9.39	4.45
1969-70	6.71	8.84	5.56
1970-71	8.86	7.82	8.14
1971-72	7.67	11.40	3.44
1972-73	8.42	11.41	3.13
1973-74	5.65	10.79	2.73
1974-75	9.56	11.25	8.29
1975-76	12.85	9.17	18.96
1976-77	9.97	11.73	17.35
1977-78	11.10	10.18+	17.16
1978-79	13.84	11.66+	17.52
1979-80	11.17	14.99+	11.74
1980-81	12.98	13.01+	11.49
1981-82	15.42	14.76+	12.77
1982-83	15.57	16.21+	15.39
1983-84	18.72	13.33+	22.45
1984-85	20.12	15.80+	28.65
1985-86	19.69	17.60+	28.38
1986-87	15.72	18.37+	23.36

Note : + includes quantities released for Food for work programme.
* Total refers to total of Central and State Governments.

Source: 1. GOI., *Bulletin on Food Statistics 1985*, Delhi.
2. GOI., *Economic Survey* 1988-89, Delhi.
3. RBI., *Report on Curency and Finance 1988-89*, Bombay.

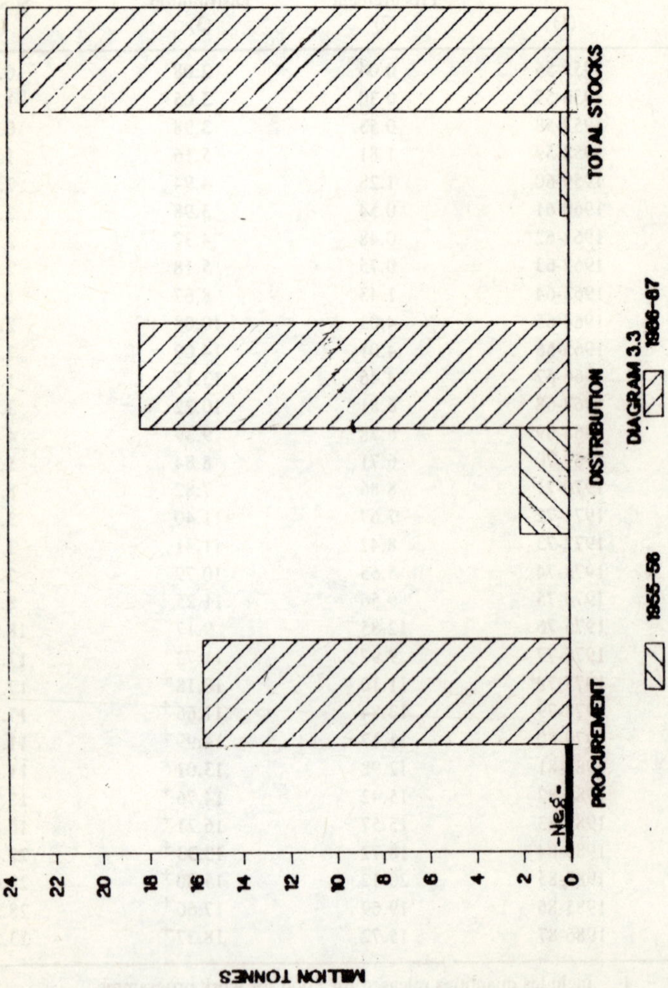

Diagram 3.3 : Procurement, Public Distribution and Stocks of Foodgrains of India

Table 3.16 : Growth Rates of Procurement, Public Distribution and Stocks of Foodgrains in India

Period	Total Procurement		Total Distribution		Total Stocks	
	Gr.	CV	Gr.	CV	Gr.	CV
1955-56 to 1966-67	30.24* (4.43)	98.11	15.18* (8.13)	60.64	10.17** (2.32)	43.96
1967-68 to 1986-87	6.13* (9.93)	38.40	3.60* (8.02)	24.54	10.79* (6.31)	63.11
1955-56 to 1986-87	14.09* (10.23)	77.71	5.03* (9.37)	43.22	11.58* (13.15)	96.89

Notes : Gr. = Exponential growth rate.
CV = Coefficient of Variation.
Figures in brackets represent 't' values.
* Significant at 1 per cent level.
** Significant at 5 per cent level.

obtained through internal procurement as well as imports. Arrangements have been made for substantial procurement of foodgrains to be supplemented by imports in order to maintain a reasonable level of public distribution, but more important, to build up buffer stocks.[28]

Trends in the public distribution of foodgrains in India are shown in Col.(3) of Table 3.15. The estimated exponential growth rates are presented in Table 3.16. It is evident from the Tables that the off-take of foodgrains from PDS increased from 2.08 million tonnes in 1955-56 to 18.37 million tonnes in 1986-87, (as shown in Diagram 3.3), *i.e.*, at a rate of 5.03 per cent annum.

The building up and maintenance of buffer stocks is an important feature of the national food policy, primarily to ensure stability in supplies and prices over the years as well as to meet contingencies arising from drought and floods. For building up a buffer stock, it is necessary that the Government itself procures the foodgrains. A new agency for procurement and building up of buffer stock, the Food Corporation of India (FCI) was established on the recommendation of

28. Investments in buffer stocks will contribute to development by providing insurance against the effects of adverse weather conditions. See. GOI., *Economic Survey 1967-68*, Delhi, p.1.

the Jha Committee. The FCI which came into existence on January 1, 1965, is to be principal agency for procurement and distribution.

Trends in the stocks of foodgrains in India are shown in Co.(4) of Table 3.15. The coefficients of variation and the estimated exponential growth rates are shown in Table 3.16. It evident from the Tables that the level of year-end closing stocks with the State and Central Governments increased from 0.32 million tonnes in 1955-56 to 1.96 million tonnes in 1966-67 in the pre-Green Revolution period, *i.e.*, at a rate of 10.17 per cent per annum. It increased further from 3.99 million tonnes in 1967-68 to 23.36 million tonnes in 1986-87, in the post-Green Revolution period, *i.e.*, at a rate of 10.79 per cent per annum. On the whole, the level of the year-end closing stocks increased from 0.32 million tonnes in 1955-56 to 23.36 million tonnes in 1986-87, (as shown in Diagram 3.3), *i.e.*, at a rate of 11.58 per cent per annum.

Food stocks reached record level in 1984-85 and 1985-86. The total size of the stocks of foodgrains (with the Centre and the States) in 1986-87 was at a comfortably high level of 23.36 million tonnes, which is a reflection of the longer turn growth in foodgrains production. Large food stocks are an important factor in lowering inflationary expectations on food prices and combating speculation in private foodgrains trade.

The coefficient of variation is found to be high for procurement, public distribution and stocks of foodgrains, 77.71 per cent, 43.22 per cent and 96.89 per cent respectively during the period, 1955-56 to 1986-87.

It is observed that, since the inception of planning, the position of the closing stocks of foodgrains has been increasing with some fluctuations. It is also noted that in spite of recent announcements that imports will be mainly utilised for building up stocks, a good part of them have been diverted for public distribution. In stepping up the stocks to the required level, there are difficulties of limited storage capacity and transport facilities.

3.9. Targets and Shortfalls in Foodgrain Production in India

The biggest problem faced by India since Independence, in fact, the entire period following World War II, is that of growing food shortage.[29] On the eve of the First Plan (1951-56) agriculture was in a deplorable condition. The Planning Commission accorded the highest priority to agriculture as the country was faced with serious food crisis and acute shortage of essential raw materials.[30] The target of production of foodgrains was set at 61.6 million tonnes. But the actual production of foodgrains by 1955-56 was at a much higher level of 69.3 million tonnes. The increase in production helped to curtail the imports of foodgrains and to control inflation and stabilize the economy.

In the Second Five-Year Plan (1951-61), the target for foodgrains production was fixed at 80.5 million tonnes,[31] while production of foodgrains touched a level of 82.3 million tonnes by 1960-61. Increase in demand for foodgrains was matched by increase in production to a large extent. As a result, the supply gap during the Second Plan period was of the small order.[32]

Achieving self-sufficiency in food was the major objective of the Third Five-Year Plan (1961-66). However, the target of foodgrain production of 100 million tonnes could not be achieved due to intermittent wars with neighbouring countries.[33] The actual production achieved by 1965-66 was 72.3 million tonnes only. The Fourth Five-Year Plan could not be introduced in April 1966 and the Government started annual planning process for next three years (1966-69). However, this period may be considered as a landmark in Indian agriculture, since the New Strategy was introduced at this time, and the embryo of the 'Green Revolution', was conceived.

The target of production for foodgrains was set at 129 million tonnes for the Fourth Plan period (1969-74). The planning Commission adopted a 'yield increase strategy' through the introduction of intensive agricultural production techniques.[34] But the targets of production could not be realised and the actual production was only 104.7 million tonnes by 1973-74.

---

29. S.K. Chatterjee, *Population and Food; Indian Economy since Independence Series*, National Publishers, Calcutta 1969, pp.31-32.
30. R. Dutt and K.P.M. Sundaram, *Indian Economy*, S. Chand and Company Ltd., New Delhi, 1986, pp.490-500.
31. S.N. Kulshreshtha and D.D. Tiwari, "Agriculture in India Problems and Prospects", *op. cit.*, pp. 111-113.
32. S.S. Madalgi, *Population and Food Supply*, *op. cit.*, p.22.
33. S.N. Kulshreshtha and D.D. Tiwari, "Agriculture in India Problems and Prospects", *op. cit.*, pp. 111-113.
34. *Ibid.*,

The target of production for foodgrains in the Fifth Plan (1974-79) was 140 million tonnes.[35] More emphasis was given to the agricultural sector. But the actual production at 131.9 million tonnes by 1978-79 fell short of the target.

The Sixth Plan (1980-85) was expected to gear the third phase of development in Indian agriculture. The Plan targeted for 154 million tonnes of foodgrain production.[36] But the actual production of foodgrains by 1984-85 is found to be 145.5 million tonnes only.

The target of production for foodgrains envisaged in the Seventh Plan (1985-90) is 183 million tonnes and the future projection of the production of foodgrains by the year 2000 A.D. has been estimated around 240 million tonnes.[37] The achievement of the target of production for foodgrains will be made possible by the increased use of fertilisers and irrigated area and by improvement in technology. Table 3.17 summarises targets and actual production of foodgrains in the Seventh Five-Year Plans of India. The targets and actual production of foodgrains are also shown in Diagram 3.4.

It is evident from Table 3.17, that there are wide fluctuations and divergencies between the targets and achievement of production of foodgrains, in the successive Five-Year Plans of India. In this context, an attempt is made in the following Chapter to examine the trends in the imports of foodgrains of India.

3.10 Summary

Foodgrains constitute the major dietary intake of the people of the Indian economy. India accounted for 10.67 per cent of the total World cereal production in 1984. While the area under total cereals in India is relatively high, the yield per hectare is found to be relatively low compared to the other major cereal producing countries of the World.

The area under foodgrains in India increased from 110.56 million hectares in 1955-56 to 127.20 million hectares in 1986-87, *i.e.*, at a rate of 0.49 per cent per annum. Production of foodgrains in India increased from 69.3 million tonnes in 1955-56 to 143.4 million

35. *Ibid.*,
36. *Ibid.*,
37. GOI., *Seventh Five-Year Plan (1985-90)*, Vol. I, Planning Commission, November 1985, pp.13-14.

Table 3.17 : Targets and Actual Production of Foodgrains During Various Five-Year Plans of India

Period of Plan	Units	Target	Actual Production	Action Programmes
I Plan_1955-56	Million Tonnes	61.6	69.3	—
II Plan_1960-61	"	80.5	82.3	—
III Plan_1965-66	"	100.0	72.3	(1) New Agricultural Strategy (2) Green Revolution
Annual Plans 1968-69	"	102.0	94.0	(3) High Yielding Varieties Programme (HYVP) (4) Intensive Agricultural District Programme (IADP) (5) Package Programmes for Cotton, Jute and Oilseeds.
IV Plan_1973-74	"	129.0	104.7	(6) Yield Increase Strategy
V Plan_1978-79	"	140.0	131.9	—
VI Plan_1984-85	"	154.0	145.5	(7) National Food Security System
VII Plan_1989-90	"	183.0	—	(8) Special Rice Production Programme in the Eastern Region. (9) National Oilseeds Development Project (10) National Watershed Development Programme for Rainfed Agriculture
1991_2000 A.D.	"	240.0	—	(11) Development of Small and Marginal Farmers (12) Social Forestry

Source: (1) *Plan documents*, Published by GOI., Planning Commission, New Delhi, Various issues
(2) GOI., *Economy Survey*, various issues.

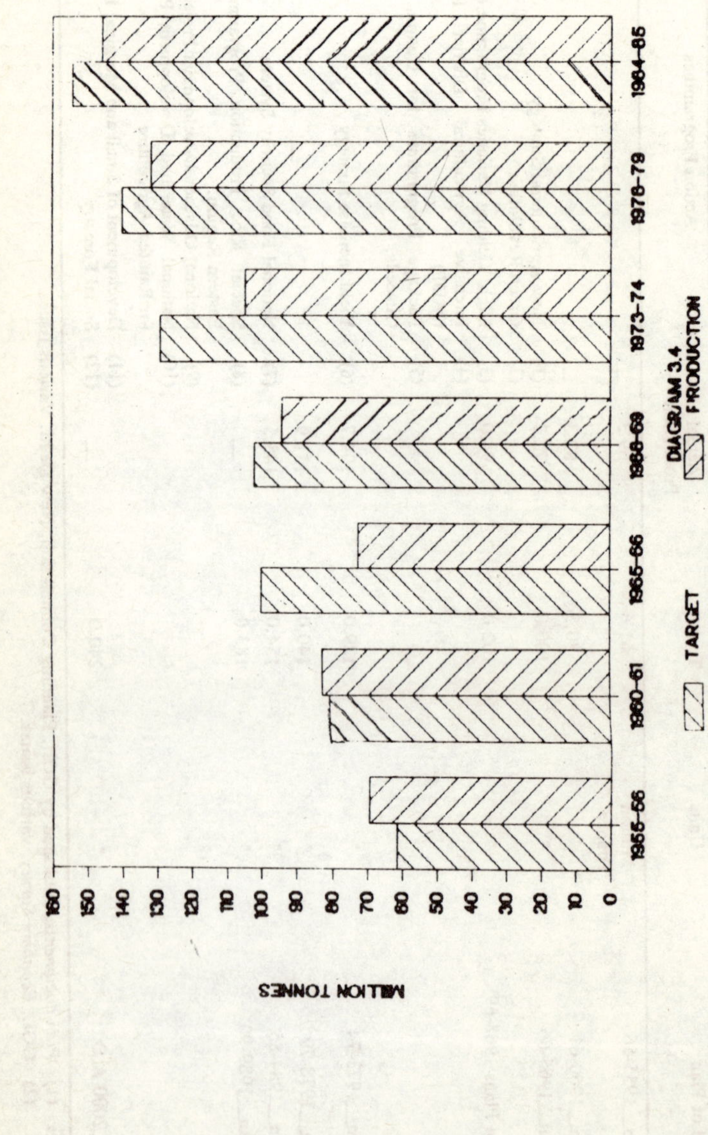

Diagram 3.4: Plan Targets and Actual Production of Foodgrains in India

tonnes in 1986-87, *i.e.*, at a rate of 2.55 per cent per annum. The peak level of 152.4 million tonnes of production of foodgrains was reached in 1983-84. Yield per hectare of foodgrains in India increased from 605 kg in 1955-56 to 1.128 kg in 1986-87, *i.e.*, at a rate of 2.13 per cent per annum, with a peak level of 1,175 kg in 1985-86. From this it may be inferred that increase in the production of foodgrains is mostly associated with increase in the yield per hectare of foodgrains, during the period 1955-56 to 1986-87.

Wide fluctuations in agricultural production have become a striking element of the Indian economy. Production of foodgrains in India showed considerable fluctuations, which emanate from fluctuations in area and yield. The coefficients of variation are found to be 25.12 per cent for production of foodgrains, 4.95 per cent for area and 21.11 per cent for yield, during the period 1955-56 to 1986-87. Thus the instability in the production of foodgrains in India seems to be mostly arising from instability in the yield per hectare of foodgrains.

Changes have taken place to some extent in the cropping pattern of foodgrains in India, with area under rice, wheat and maize showing an increase and area under jowar, small millets and barley showing a decline, during the period 1955-56 to 1986-87. Rice and wheat together accounted for 50.7 per cent of the total area under foodgrains in 1986-87.

Rice and wheat together accounted for 73.15 per cent of the total foodgrains production in 1986-87. There are some changes in the pattern of foodgrains production in India during the period 1955-56 to 1986-87. There is a rise in the relative share of rice, wheat and maize, while there is decline in the relative share of jowar, bajra, ragi, barley and small millets. Thus, there is a shift towards the production of superior grains from coarse-grains and towards greater production of cereals compared to pulses.

The index number of wholesale prices of total foodgrains in India increased from 35.20 in 1955-56 (with 1970-71 = 100) to 298.60 in 1986-87, *i.e.*, at a rate of 6.79 per cent per annum. Price of cereals, pulses, and foodgrains have not only shown an increasing trend, but also considerable fluctuations. There is a sharp rise in the prices of the foodgrains due to various factors like drought and flood conditions, increase in population, urbanisation and increase in incomes of the people.

The per capita net availability of foodgrains in India increased only marginally, at a rate of 0.24 per cent per annum during the period 1955-56 to 1986-87, with a peak level of 480.2 grams per day in 1964-45. The level of per capita net availability of foodgrains per day is found to be lower than the minimum nutrition requirements of 594 grams per day, as specified by Indian Council of Medical Research (ICMR).

Arrangements have been made for substantial procurement of foodgrains to be supplemented by imports in order to maintain a reasonable level of public distribution but more important, to build up buffer stocks. Foodgrains procured by the Government increased from 0.04 million tonnes in 1955-56 to 15.72 million tonnes in 1986-87, *i.e.*, at a rate of 14.09 per cent per annum. During this period, the off-take of foodgrains from public distribution system (PDS) increased from 2.08 million tonnes in 1955-56 to 18.37 million tonnes in 1986-87, *i.e.*, at a rate of 5.03 per cent annum. The building up and maintenance of buffer stocks is an important feature of the national food policy, primarily to ensure stability in supplies and price over the years as well as to meet contingencies arising from drought and floods. The level of the year-end closing stocks of foodgrains increased from 0.32 million tonnes in 1955-56 to 23.36 million tonnes in 1986-87, *i.e.*, at a rate of 11.58 per cent per annum.

Targets have been set for foodgrains production under successive Five-Year Plans. The analysis of targets and actual production of foodgrains in Seven Five-Year Plans of India indicates that there are wide fluctuations and divergencies between the targets and achievement of production of foodgrains in the successive Plans of India, except in the First Five-Year Plan.

In the context of the need for better food management, trends in the imports of foodgrains in India and their role in the Indian economy are examined in the next Chapter.

CHAPTER 4

TRENDS IN THE IMPORTS OF FOODGRAINS IN INDIA

FOOD Problem is one of the important problems which India has been facing. With the production of foodgrains not being adequate, the deficiency in the requirement of foodgrains is being met through imports of foodgrains. In this context, an attempt is made in this Chapter to examine trends in the imports of food and non-food items, trends in the imports of foodgrains the pattern of imports of foodgrains, the direction of imports of foodgrains and the degree of dependency of the Indian economy on foodgrain imports.

4.1. Structure of Imports of India

Before proceeding to analyse the trends in the imports of food non-food items, an attempt is made here to analyse the structure of imports of India, in order to bring out the importance of 'Food' in the imports of India.

Data relating to the percentage share of some of the major commodities imported by India in the total imports of India, during the past eight years, *i.e.*, 1980-81 to 1987-88, are shown in Table 4.1. It is evident from Table that 'petroleum and petroleum products' accounted for 41.90 per cent of the total imports of India in 1980-81, but declined to 18.23 per cent in 1987-88. Imports of 'capital goods' constitute another major item in the import structure of India, accounting for 15.20 per cent of the total imports in 1980-81. The relative share of imports of capital goods increased to 27.06 per cent in 1986-87, due to the buoyancy in the climate for industrial investment and on account of the modernisation and upgradation of technology.[1] Capital goods accounted for 28.06 per cent of the total

1. GOI., *Economic Survey 1987-88*, Delhi, p.86.

Table 4.1 : Structure of Imports of India

(Percentage Distribution)

	Commodity (1)	1980-81 (2)	1981-82 (3)	1982-83 (4)	1983-84 (5)	1984-85 (6)	1985-86 (7)	1986-87 (8)	1987-88P (9)
1.	Petroleum and Petroleum Products	41.90	38.10	39.30	30.51	31.41	27.50	24.70	18.23
2.	Capital goods	15.20	12.80	19.00	20.98	18.49	21.80	27.06	28.06
3.	Iron and Steel	6.20	8.35	8.00	6.08	5.49	7.09	7.18	5.68
4.	Fertilisers finished	5.19	3.75	1.43	0.71	5.88	5.36	2.45	0.77
5.	Edible Oils	5.01	2.78	1.58	3.42	5.54	3.70	3.03	4.10
6.	Non-ferrous metals	3.38	2.00	1.95	2.33	2.40	2.76	2.05	2.57
7.	Pearls, precious and semi precious stones	3.32	2.54	4.70	6.84	6.02	5.59	7.40	8.90
8.	Organic and inorganic chemicals	2.59	2.85	2.70	4.17	3.42	5.50	5.13	4.69
9.	Paper-paper boards and manufactures thereof	1.49	1.77	1.00	1.09	1.14	1.14	0.96	1.15

10.	Professional and scientific instruments etc.	1.24	1.10	1.00	1.77	1.68	1.90	2.26	2.19
11.	Non-metalic, mineral manufactures (excluding pearls)	1.20	0.84	0.99	1.10	0.48	0.50	0.42	0.37
12.	Cereals and cereal preparations	0.80	2.55	2.60	5.10	1.41	0.56	0.23	0.15
13.	Synthetic and re-generated fibres	0.77	1.20	0.80	0.66	0.35	0.35	0.22	0.13
14.	Medicinal and Pharmaceuticals	0.67	0.53	0.56	0.83	0.80	0.90	0.78	5.68
15.	Chemical materials and products	0.54	0.40	0.50	0.78	0.79	0.70	0.59	0.78
16.	Fertiliser crude	0.50	0.59	1.00	0.50	0.80	0.80	0.65	0.62
17.	Others	10.00	17.85	12.89	13.13	13.90	13.84	14.88	15.91
Total		100.00	100.00	100.00	100.00	100.00	100.00	100.00	100.00

Note: (P) Privisional.
Source: Based on GOI., *Economic Survey 1983-84 and 1988-89*, Delhi.

imports in 1987-88. Imports of iron and steel accounted for 6.20 per cent of the total imports, followed by imports of finished fertilisers with 5.19 per cent and imports of edible oils with 5.01 per cent of the total imports of India in 1980-81.

Imports of 'Cereals and Cereal preparations' accounted for 0.80 per cent of the total imports of India in 1980-81. The relative share of imports of Cereals increased to 5.10 per cent in 1983-84 but declined to 0.15 per cent in 1987-88.

These various commodities imported by India may broadly be classified into four major groups, *viz.*, (*a*) Food and Live Animals Chiefly for Food, (*b*) Raw Materials and Intermediate Manufacturers, (*c*) Capital Goods and (*d*) others (unclassified). Imports of Cereals and Cereal preparations (Cereals for short) come under the Group of 'Food and Live Animals Chiefly for Food'.

The data relating to the principal imports of India, classified into the above four major groups for the period 1955-56 to 1985-86 are presented in Table 4.2. From the Table, it is evident that the value of imports of the commodity group 'Food and Live Animals Chiefly for Food', increased from Rs. 57.76 crores in 1955-56 to Rs. 853.70 crores in 1985-86. However, its share in the total imports of India declined from 7.46 per cent in 1955-56 to 4.34 per cent in 1985-86. It may also be noted from the Table that the percentage share of imports of Cereals and Cereal preparations in the Group of 'Food and Live Animals Chiefly for Food', increased from 60.75 per cent in 1955-56 to 87.87 per cent in 1970-71, but declined to 12.89 per cent in 1985-86.

The relative share of imports of the commodity group 'Raw Materials and Intermediate Manufactures', increased from 49.98 per cent in 1955-56 to 71.05 per cent in 1985-86. The relative share, of the imports of the third group *viz.*, 'Capital goods', however, declined from 38.10 per cent in 1955-56 to 21.80 per cent in 1985-86. The relative share of the value of imports of the fourth group, *viz.*, 'Others', (unclassified) declined from 4.46 per cent in 1955-56 to 2.81 per cent in 1985-86, of the total imports of India.

Thus, there is some change in the structure of Imports of India, with the relative share of the Group 'Raw Materials and intermediate manufactures' increasing over time and that of 'Food' imports showing a decline.

Table 4.2 : Principal Imports of India : By Commodity Groups

(Value : Rs. crores)

S.No.	Commodity Group	1955-56	Share of Commodity group in total imports	1970-71	Share of Commodity group in total imports	1980-81	Share of Commodity group in total imports	1985-86	Share of Commodity group in total imports
1.	Food and Live Animals, Chiefly for Food (excluding cashew raw)	57.76 (100.00)	7.46	242.40	14.83	380.20	3.03	853.70	4.34
	of which: a) Cereals and Cereal Preparations*	35.09 (60.75)	—	213.00 (87.87)	—	100.40 (26.41)	—	110.00 (12.89)	—
2.	Raw materials and Intermediate manufactures	386.85	49.98	888.60	54.38	9,759.60	77.77	13,966.10	71.05
3.	Capital goods	294.92	38.10	404.00	24.72	1,910.30	15.20	4,285.40	21.80
4.	Others (unclassified)	34.47	4.46	99.20	6.07	499.10	3.98	552.50	2.81
5.	Total Imports of India	774.00	100.00	1,634.20	100.00	12,594.20	100.00	19,657.70	100.00

Note: Figures in brackets indicate percentage share of Cereals and Cereal preparations in the Group 'food and Live Animals, Chiefly for Food'.

Source: Based on GOI., *Economy Survey, 1960-61 and 1988-89*, Delhi.

The above analysis brings out the relative importance of imports of Cereals among the principal imports of India in general, and among the imports of 'Food and Live Animals Chiefly for Food', in particular.

4.2. Trends in Food and Non-Food Imports of India

Table 4.3 gives percentage distribution of India's imports in terms of Food and Non-Food items during the period, 1955-56 to 1987-88. In order to give a reasonably realistic picture, imports of Food items, Non-Food items (excluding manufactured fertilisers and petroleum products), Manufactured Fertilisers and Petroleum Products are considered separately. The relative share of each of these items in the total imports of India are presented in the Table for purposes of comparison.

It is evident from Table 4.3 that the share of Food imports increased from 7.46 per cent in 1955-56 to a peak level of 38.20 per cent in 1966-67, but declined to 4.34 per cent in 1985-86 in the total imports of India. Similarly, the relative share of Non-food imports (excluding manufactured fertilisers and petroleum products) declined from 91.53 per cent in 1955-56 to 62.79 per cent in 1985-86. On the other hand, the relative share of imports of manufactured fertilisers given in Col.(4) of the Table increased from 0.29 per cent in 1955-56 to 5.36 per cent in 1985-86, touching a peak level of 9.66 per cent in 1974-75. The relative share of imports of petroleum and petroleum products, given in Col. (5) of the Table, increased[2] from 0.72 per cent in 1955-56 to 27.50 per cent in 1985-86.

On the whole, during the period, 1985-86, the relative share of imports of Food items showed a decline. The relative share of Non-Food items also showed a decline, while the relative shares of Manufactured Fertilisers and Petroleum Products, in the total Imports of India, showed increase during the same period.

2. The steep rise in the share of petroleum and petroleum products is largely the result of the increase in the price of crude oil in the International markets. The relative share is found to be 18.23 per cent in 1987-88.

Table 4.3 : Percentage Distribution of India's Imports of Food and Non-Food Items: 1955-56 to 1987-88

Year	Food Imports	Non-Food Imports (excluding manufactured fertilisers and petroleum products)	Manufactured Fertilisers	Petroleum and Petroleum Products	Total Imports
(1)	(2)	(3)	(4)	(5)	(6)
1955-56	7.46	91.53	0.29	0.72	100.00
1956-57	6.22	92.50	0.44	0.84	100.00
1957-58	15.83	82.71	0.41	1.05	100.00
1958-59	20.67	77.50	0.95	0.88	100.00
1959-60	18.92	71.29	1.64	8.15	100.00
1960-61	26.16	66.47	1.01	6.35	100.00
1961-62	14.50	74.88	1.20	9.42	100.00
1962-63	15.86	74.02	2.36	7.76	100.00
1963-64	18.74	69.12	3.01	9.13	100.00
1964-65	20.91	71.86	2.15	5.08	100.00
1965-66	22.85	69.54	7.77	4.84	100.00
1966-67	38.20	54.08	4.50	3.22	100.00
1967-68	29.37	60.02	6.82	3.79	100.00
1968-69	21.11	64.60	7.29	7.00	100.00
1969-70	20.27	66.16	4.87	8.70	100.00

(Contd.)

Table 4.3 (Cont.)

(1)	(2)	(3)	(4)	(5)	(6)
1970-71	14.83	85.17	3.75	8.31	100.00
1971-72	10.87	73.94	4.48	10.71	100.00
1972-73	8.56	75.89	8.56	1.93	100.00
1973-74	18.51	57.02	5.51	18.96	100.00
1974-75	18.92	45.82	9.66	25.60	100.00
1975-76	26.50	39.96	8.98	23.76	100.00
1976-77	19.22	48.68	3.94	28.16	100.00
1977-78	3.86	66.09	4.29	25.76	100.00
1978-79	3.75	72.32	5.46	18.47	100.00
1979-80	3.06	57.14	4.06	35.74	100.00
1980-81	3.03	49.87	5.19	41.90	100.00
1981-82	5.07	53.05	3.75	38.10	100.00
1982-83	4.46	54.81	1.43	39.30	100.00
1983-84	6.43	62.35	0.71	30.51	100.00
1984-85	4.06	58.66	5.88	31.41	100.00
1985-86	4.34	62.79	5.36	27.50	100.00
1986-87	NA	NA	2.45	24.70	100.00
1987-88(P)	NA	NA	0.77	18.23	100.00

Note: P. = Provisional

Source:
1. GOI., *Statistical Abstract: India*, various issues, Delhi.
2. RBI., *Report on Currency and Finance*, various issues, Bombay.
3. GOI., *Economic Survey*, various issues, Delhi.

4.3. Trends in the Value of Imports of Cereals and Cereal Preparations of India

As pointed out earlier, imports of foodgrains come under the category of imports of 'Food and Live Animals Chiefly for Food'. In this Chapter, imports of foodgrains are represented by the imports of Cereals and Cereal Preparations (Cereals in short). Trends in the value of imports of Cereals, imports of 'Food and Live Animals Chiefly for Food', and total Imports of India are presented in Columns (2), (4) and (6) of Table 4.4., for the period 1955-56 to 1986-87 (the latest year for which comparable data are available). Three-year moving averages have been computed for imports of Cereals, imports of Food and total Imports of India. The three-year moving averages are shown in Columns (3), (5) and (7) of Table 4.4. The relative shares of each of these imports in the total imports of India are shown in Table 4.5 for purposes of comparison. Exponential growth rates[3] have been estimated for imports of Cereals, imports of Food and total Imports of India. The coefficients of variation and estimated exponential growth rates are presented in Table 4.6.

From Tables 4.4, 4.5 and 4.6, it is evident that imports of Cereals in terms of value increased from Rs. 35.09 crores in 1955-56 to Rs. 192.90 crores in 1960-61 and further to Rs. 266.30 crores in 1964-65. There is a steep increase in the value of imports of Cereals from Rs. 266.30 crores in 1964-65 to Rs. 322.00 crores in 1965-66, mainly due to the severe drought which affected the Indian economy in 1965-66. The drought prevailed in India in 1966-67 also for the second year in succession.

The trend in Cereal imports during 1965-66 and 1966-67 reflects of strain imposed on the economy by the two successive droughts. There has been a substantial shift in composition in favour of emergency Food imports, (as shown in Col. (2) of Table 4.3) at the expense of intermediate and capital goods.[4] The relative share of foodgrain imports in total imports increased from 4.53 per cent in 1955-56 to 22.85 per cent in 1965-66, *i.e.*, the last year of the Third Five-Year Plan.

---

3. Exponential growth rates are estimated by employing the formula as given in Chapter 2; $Y = ae^{bt}$, where Y = imports, a = constant, b = growth rate and t = time period.
4. GOI., *Economic Survey 1966-67*, Delhi, p.47.

Table 4.4 : Trends in the Value of Imports of Cereals and Cereal Preparations of India

(Value; Rs. crores)

Year	Imports of Cereals and Cereals preparations	3-year moving average	Imports of Food and Live Animals Chiefly for Food	3-year moving average	Total Imports of India	3-year moving average
(1)	(2)	(3)	(4)	(5)	(6)	(7)
1955-56	35.09	—	57.76	—	774	—
1956-57	56.30	84.60	56.30	92.15	906	902
1957-58	162.40	113.07	162.40	132.44	1,026	932
1958-59	120.50	141.43	178.61	173.85	864	950
1959-60	141.40	151.37	181.54	215.28	960	972
1960-61	192.90	154.40	285.70	204.77	1,902	1,022
1961-62	129.60	154.30	147.07	204.6	1,015	1,079
1962-63	141.10	151.43	179.41	180.29	1,131	1,097
1963-64	183.60	197.00	214.40	225.30	1,144	1,208
1964-65	266.30	257.30	282.10	272.83	1,349	1,301
1965-66	322.00	370.47	322.00	418.37	1,409	1,487
1966-67	523.10	459.10	651.00	517.38	1,704	1,695
1967-68	532.20	463.97	597.14	544.40	1,972	1,862
1968-69	336.60	373.93	403.06	434.30	1,909	1,821
1969-70	253.00	267.53	320.70	322.05	1,582	1,708

1970-71	213.00	199.07	242.40	253.34	1,634	1,676
1971-72	131.20	141.67	196.93	199.69	1,812	1,771
1972-73	80.80	228.40	159.75	301.25	1,867	2,211
1973-74	473.20	439.27	457.07	520.67	2.955	3,114
1974-75	763.80	859.93	855.20	932.56	4,519	4,246
1975-76	1,342.80	991.40	1,395.40	1,039.40	5,265	4,933
1976-77	867.60	777.63	937.40	831.82	5,074	5,433
1977-78	122.50	359.00	214.50	448.60	6,020	5,941
1978-79	86.90	105.07	245.72	252.70	6,811	7,317
1979-80	105.80	97.70	279.91	301.94	9,143	9,949
1980-81	100.40	184.40	380.20	450.07	12,549	11,766
1981-82	347.00	273.53	690.10	569.50	13,608	13,483
1982-83	373.20	509.73	638.20	782.13	14,293	14,578
1983-84	809.00	474.73	1,018.10	783.70	15,832	15,753
1984-85	242.00	387.00	694.80	855.53	17,134	17,541
1985-86	110.00	133.00	853.70	—	19,658	18,998
1986-87	47.00	—	NA	—	20,201	—

Source: 1. GOI., *Economy Survey*, various issues, Delhi.
2. GOI., Ministry of Commerce and Industry, *Annual Administration of the Import and Export Trade Control* 1962, Delhi.
3. GOI., CSO., *Statistical Abstract India*, various issues, Delhi.
4. RBI., *Report on Currency and Finance*, various issues, Bombay.

Table 4.5 : Relative Share of Imports of Cereals and Cereal Preparations of India

(Percentage)

Period	Share of Cereals and Cereal Preparations in Food and Live Animals Chiefly for Food	Share of Cereals and Cereal Preparations in Total Imports of India	Share of Food and Live Animal Imports in Total Imports of India	Total Imports of India
(1)	(2)	(3)	(4)	(5)
1955-56	60.75	4.53	7.46	100.00
1956-57	100.00	6.21	6.22	100.00
1957-58	100.00	15.83	15.83	100.00
1958-59	67.47	13.95	20.67	100.00
1959-60	77.89	14.73	18.92	100.00
1960-61	67.52	17.66	26.16	100.00
1961-62	88.12	12.77	14.50	100.00
1962-63	78.65	12.48	15.86	100.00
1963-64	85.63	16.05	18.74	100.00
1964-65	94.40	19.74	20.91	100.00
1965-66	100.00	22.85	22.85	100.00
1966-67	80.35	30.70	38.20	100.00
1967-68	91.89	26.99	29.37	100.00
1968-69	83.51	17.63	21.11	100.00

1969-70	78.89	15.99	20.27	100.00
1970-71	87.87	13.03	14.83	100.00
1971-72	66.62	7.24	10.87	100.00
1972-73	50.58	4.33	8.56	100.00
1973-74	86.50	16.01	18.51	100.00
1974-75	89.31	16.90	18.92	100.00
1975-76	96.23	25.50	26.50	100.00
1976-77	100.00	17.30	19.22	100.00
1977-78	52.69	2.03	3.86	100.00
1978-79	35.37	1.28	3.75	100.00
1979-80	37.80	1.16	3.06	100.00
1980-81	26.41	0.80	3.03	100.00
1981-82	50.28	2.55	5.07	100.00
1982-83	58.48	2.60	4.46	100.00
1983-84	79.46	5.10	6.43	100.00
1984-85	34.83	1.41	4.06	100.00
1985-86	12.89	0.56	4.34	100.00
1986-87	N.A	0.23	N.A	100.00

Source: Based on Table 4.4.

Table 4.6 : Growth Rates of Cereal Imports, Food Imports and Total Imports of India

Period	Cereal Imports Gr.	Imports of Food and Live Animals Gr.	Total Imports Gr.
1955-56 to 1966-67	17.69 * (5.97)	16.23 * (5.27)	5.73 * (7.91)
1967-68 to 1986-87[a]	-4.12 (1.14)	4.40 ** (1.74)	15.36 * 19.11)
1955-56 to 1986-87	2.02 (1.20)	6.20 * (5.48)	11.30 * (20.12)

Note: (a) 1985-86 for Imports of Food and Live Animals.
Gr. - Exponential growth rate.
Figures in brackets refer 't' values.
* Significant at one per cent level
** Significant at ten per cent level.

Foodgrain imports continued to increase sharply in terms of value from Rs. 322.00 crores in 1965-66 to Rs. 523.10 crores in 1966-67. On the whole, imports of Cereals increased from Rs. 35.09 crores in 1955-56 to Rs. 523.10 crores in 1966-67, *i.e.*, at a rate of 17.69 per annum during the pre-Green Revolution period. Imports of Cereals rose further to Rs. 532.20 crores in 1967-68, in terms of value.

Value of Cereal imports showed a declining trend from Rs. 532.20 crores in 1967-68 to Rs.80.80 crores in 1972-73. With the increase in domestic production, reliance on imports of foodgrains has tended to diminish.[5] In 1973-74, imports of Cereals increased sharply to Rs. 473.20 crores, and rose further to Rs. 763.80 crores in 1974-75 and to a peak level of Rs. 1.342.80 crores in 1975-76. The relative share of imports of Cereals in the total imports of India increased from 16.01 per cent in 1973-74 to 25.50 per cent in 1975-76.

Thus, Cereals became the largest single item of imports in 1975-76 and accounted for more than a quarter of the total value of imports of India, exceeding[6] the relative share of even the value of imports of

5. GOI., *Economic Survey 1969-70*, Delhi, p.44.
6. Larger quantity of imports and an increase of 23.2 per cent in the unit value contributed to the higher import-bill. See, GOI., *Economic Survey 1976-77*, Delhi, p.40.

POL. The value of Cereal imports continued to be at a high level of Rs. 867.60 crores in 1976-77, but declined to Rs. 122.50 crores in 1977-78. The relative share of imports of Cereals in the total imports of India declined from 17.30 per cent in 1976-77 to 2.03 per cent in 1977-78. Imports of Cereals declined further to Rs. 86.90 crores in 1978-79, but rose to Rs. 105.80 crores in 1979-80.

Imports of Cereals increased from Rs. 100.40 crores in 1980-81 to Rs. 347.00 crores in 1981-82. Cereal imports rose further to Rs. 373.20 crores in 1982-83 and reached a high level of Rs. 809.00 crores in 1983-84, but declined to a low level of Rs. 47.00 crores in 1986-87. Thus, imports of Cereals declined from Rs.532.20 crores in 1967-68 to Rs. 47.00 crores in 1986-87, *i.e.*, at a rate of 4.12 per cent per annum, though not significant, during the post-Green Revolution period.

On the whole, imports of Cereals of India, in terms of value, increased from Rs. 35.09 crores in 1955-56 to Rs. 47.00 crores in 1986-87, *i.e.*, at a rate of 2.02 per cent per annum, though not significant.

Col. (4) of Table 4.4. gives the imports of 'Food and Live Animals Chiefly for Food' in value terms. Imports of Food increased from Rs. 57.76 crores in 1955-56 to Rs. 651.00 crores in 1966-67, *i.e.*, at a rate of 16.23 per cent per annum in the pre-Green Revolution period. During the post-Green Revolution period, imports of Food increased from Rs. 579.14 crores in 1967-68 to Rs. 853.70 crores in 1985-86, *i.e.*, at a rate of 4.40 per cent per annum. On the whole, imports of food and Live Animals Chiefly for Food increased from Rs. 57.76 crores in 1955-56 to Rs. 853.70 crores in 1985-86, *i.e.*, at a rate of 6.20 per cent per annum. The peak level of Food imports of Rs. 1,395.40 crores was reached in the year 1975-76. Cereal imports were also at a peak level in this year.

Total Imports of India [as shown in Col. (6)] increased from Rs. 774 crores in 1955-56 to Rs. 1,704 crores in 1966-67, *i.e.*, at a rate of 5.73 per cent per annum. Total imports increased further from Rs. 1,972 crores in 1967-68 to Rs. 20,201 crores in 1986-87, *i.e.*, at a rate of 15.36 per cent per annum. On the whole, total imports of India increased from Rs. 774 crores in 1955-56 to Rs. 20,201 crores in 1986-87, *i.e.*, at a rate of 11.30 per cent per annum.

Imports of Cereals accounted for 60.75 per cent in 1955-56 of the value of imports of Food and Live Animals. In 1956-57 imports of Food and Live Animals consisted solely of imports of Cereals. This was the position in the years 1957-58, 1965-66 and 1976-77. In 1985-86, however, Cereal imports accounted for 12.89 per cent of the imports of Food and Live Animals. Thus, the relative share of imports of Cereals in the imports of Food and Live animals showed wide variations.

Import of Cereals accounted for 4.53 per cent in 1955-56 of the total Imports of India, and 30.70 per cent in 1966-67. It declined to 4.33 per cent in 1972-73, but increased again to 25.50 per cent in 1975-76, but declined again to 0.23 per cent in 1986-87. Thus, the relative share of import of Cereals in the total Imports of India also showed considerable fluctuations during the period 1955-56 to 1986-87. The relative share of imports of Food and Live Animals in the total imports of India showed a decline from 7.46 per cent in 1955-56 to 4.34 per cent in 1985-86, with a peak level of 38.20 per cent in 1966-67.

4.4 Quantum of Foodgrain Imports of India

Trends in the imports of foodgrain of India in terms of quantity are shown in Col.(2) of Table 4.7 along with the imports of foodgrains in terms of value in Col. (4) for purposes of comparison. These trends are also shown in Graph 4.1 and Graph 4.2. Three-year moving average for foodgrain imports, in terms of quantity and value, are constructed and shown in Cols. (3) and (5) of the same Table. Exponential growth rates are estimated for quantity and value of imports of foodgrains for the pre-Green Revolution period 1955-56 to 1966-67, for the post-Green Revolution period 1967-68 to 1986-87 and for the total period 1955-56 to 1986-87. The estimated exponential growth rates are shown in Table 4.8.

From Tables 4.7 and 4.8, it is evident that imports of foodgrains in terms of quantity increased from 0.71 million tonnes in 1955-56 to 5.14 million tonnes in 1960-61. This increase was mainly due to unfavourable weather conditions in the economy. Foodgrain imports declined to 3.49 million tonnes in 1961-62, but rose again from 3.64 million tonnes in 1962-63 to 4.62 million tonnes in 1963-64.

Table 4.7 : Trends in the Quantity and Value of Foodgrain Imports of India : 1955-56 to 1986-87

Q - Million tonnes
V - Rs. in crores

Period	Quantity	3-year moving average	Value	3-year moving average
(1)	(2)	(3)	(4)	(5)
1955-56	0.71	—	35.09	—
1956-57	1.42	1.94	56.30	84.60
1957-58	3.68	2.77	162.40	113.07
1958-59	3.22	3.59	120.50	141.43
1959-60	3.87	4.08	141.40	151.37
1960-61	5.14	4.17	192.20	154.40
1961-62	3.49	9.85	129.60	154.30
1962-63	3.64	3.92	141.10	151.43
1963-64	4.62	4.88	183.60	197.00
1964-65	6.70	6.16	266.30	257.30
1965-66	7.46	8.07	322.00	370.47
1966-67	10.36	8.83	523.10	459.10
1967-68	8.67	8.24	532.20	463.97
1968-69	5.69	6.08	336.60	373.93
1969-70	3.87	4.40	253.00	267.53
1970-71	3.63	3.19	213.00	199.07
1971-72	2.05	2.16	131.20	141.67
1972-73	0.80	2.16	80.80	228.40
1973-74	4.40	3.21	473.20	439.27
1974-75	5.60	5.60	763.80	859.93
1975-76	7.56	6.43	1.342.80	991.40
1976-77	6.52	4.91	867.60	777.63
1977-78	0.65	2.47	122.50	359.00
1978-79	0.24	0.43	86.90	105.07
1979-80	0.41	0.36	105.80	97.70
1980-81	0.43	0.79	100.40	184.40
1981-82	1.52	1.09	347.00	273.53
1982-83	1.34	2.36	373.20	509.73
1983-84	4.21	2.17	809.00	474.73
1984-85	0.95	1.82	242.00	387.00
1985-86	0.30	0.46	110.00	133.00
1986-87	0.13	—	47.00	—

Source:
1. GOI., Bulletin on Food Statistics, Delhi, various issues.
2. U.N., FAO., Trade Year Book, various issues, Delhi.
3. U.N., Year Book of International Trade Statistics various issues, New York.
4. GOI., Economic Survey, various issues, Delhi.
5. GOI., CSO., Statistical Abstract : India, various issues, Delhi.
6. RBI., Report on Currency and Finance, various issues, Bombay.

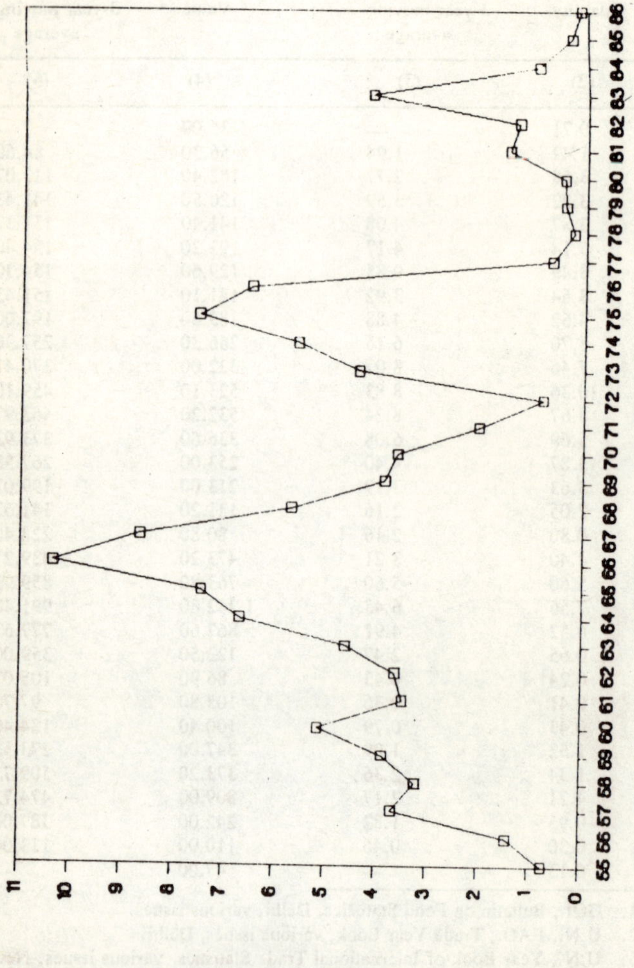

GRAPH 4.1

Graph 4.1: Trends in the Quantity of Foodgrain Imports of India

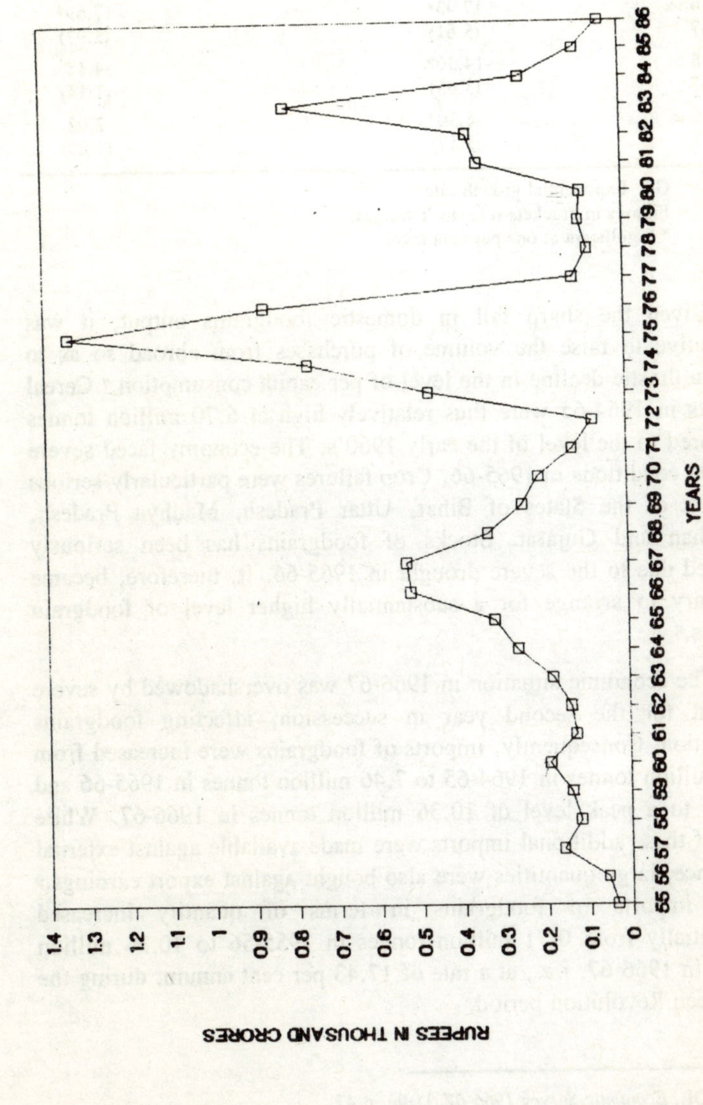

Graph 4.2: Trends in the Value of Foodgrain Imports of India

Table 4.8 : Growth Rates of Quantity and Value of Imports of Foodgrains of India

Period	Quantity Gr.	Value Gr.
1955-56 to 1966-67	17.43* (5.61)	17.69* (5.97)
1967-68 to 1986-87	-14.30* (3.65)	-4.12 (1.14)
1955-56 to 1986-87	-6.30* (3.17)	2.02 (1.20)

Note: Gr. -Exponential growth rate.
Figures in brackets refer to 't' values.
* Significant at one per cent level.

Given the sharp fall in domestic foodgrains output, it was imperative to raise the volume of purchases from abroad so as to avoid a drastic decline in the level of per capita consumption.[7] Cereal imports in 1964-65 were thus relatively high at 6.70 million tonnes compared to the level of the early 1960's. The economy faced severe drought conditions in 1965-66. Crop failures were particularly serious in parts of the States of Bihar, Uttar Pradesh, Madhya Pradesh, Rajasthan and Gujarat. Stocks of foodgrains has been seriously depleted due to the severe drought in 1965-66. It, therefore, became necessary to arrange for a substantially higher level of foodgrain imports.[8]

The economic situation in 1966-67 was overshadowed by severe drought for the second year in succession, affecting foodgrains production. Consequently, imports of foodgrains were increased from 6.70 million tonnes in 1964-65 to 7.46 million tonnes in 1965-66 and further to a peak level of 10.36 million tonnes in 1966-67. While most of these additional imports were made available against external assistance, large quantities were also bought against export earnings.[9] Thus, imports of foodgrains in terms of quantity increased substantially from 0.71 million tonnes in 1955-56 to 10.36 million tonnes in 1966-67, *i.e.*, at a rate of 17.43 per cent annum, during the pre-Green Revolution period.

7. GOI., *Economic Survey 1966-67*, Delhi, p.47.
8. *Ibid.*, p.1
9. *Ibid.*, p.2.

In the year 1966-67, Government introduced the use of New Technology popularly known as 'Green Revolution', in order to increase production in the agricultural sector. With a record harvest of 95.1 million tonnes of foodgrains in 1967-68 and the maintenance of foodgrains production at approximately the same level in 1968-69, it was not felt necessary to import foodgrains on the same scale as before.[10] Hence, imports of foodgrains declined from 10.36 million tonnes in 1966-67 to 8.67 million tonnes in 1967-68. There was further decline in foodgrain imports to 5.69 million tonnes in 1968-69. While in the earlier years, imports had to be made to sustain domestic consumption, in 1968-69 with the easing of the pressure on the public distribution system (PDS), they were also used for the building up of buffer stocks. Better internal procurement also made it easier to import smaller quantities from abroad for both these purposes.[11]

Increase in domestic production from 94 million tonnes in 1968-69 million tonnes in 1969-70 and further to 108.4 million tonnes in 1970-71, tended to reduce the reliance on imports of foodgrains. Imports of foodgrains declined from 3.87 million tones in 1969-70 to 3.63 million tonnes in 1970-71. There. was a further decline in foodgrain imports to 2.05 million tonnes in 1971-72 and 0.80 million tonnes in 1972-73.

Due to sluggishness of the domestic output in 1973-74, foodgrain imports on a substantially large scale of 4.40 million tonnes in 1973-74 and 5.60 million tonnes in 1974-75 became necessary. Imports of foodgrains touched a high level of 7.56 million tonnes in 1975-76 necessitated by the drought conditions prevailing in the economy. In 1975-76, foodgrain imports increased substantially both in terms of quantity and value. The effect of steep increase in the quantity of imported foodgrains has been compounded by the higher prices of foodgrains in the international markets due to World wide inflation in wake of a steep increase in the price of oil.[12] In 1976-77, foodgrain imports showed a decline but were still at a relatively high level of 6.52 million tonnes. Foodgrain imports declined sharply to 0.65 million tonnes in 1977-78 and 0.24 million tonnes in 1978-79, due to improved domestic production.

---

10. GOI., *Economic Survey 1969-70*, Delhi, p.44
11. *Ibid.*
12. GOI., *Economic Survey 1975-76*, Delhi, p.42

The beginning of the Sixth Five-Year Plan was marked by the drought of 1979. Imports of Cereals increased from 0.43 million tonnes in 1980-81 to 1.52 million tonnes in 1981-82. These imports were undertaken by the Government with a view to building up buffer stocks and to combat pressure on prices.[13] The DGCI & S data show that import of Cereals more than doubled and touched a high level of 4.21 million tonnes in 1983-84, as the harvest position in 1982-83 has been unsatisfactory.[14] Imports of foodgrains continued to supplement domestic availability after the drought in 1982-83.[15] The sharp reduction in foodgrain imports from 0.95 million tonnes in 1984-85 to 0.30 million tonnes in 1985-86 has been made possible by successive good harvests and comfortable stocks position of foodgrains.[16] Thus, the imports of foodgrains in terms of quantity declined from 8.67 million tonnes in 1967-68 to 0.13 million tonnes in 1986-87, *i.e.*, at a rate of 14.30 per cent per annum, during the post-Green Revolution period.

Foodgrain imports of India in terms of value increased from Rs. 35.09 crores in 1955-56 to Rs. 523.10 crores in 1966-67, *i.e.*, at a rate of 17.69 per cent per annum. This growth rate was slightly higher than the growth rate of foodgrain imports in terms of quantity during the pre-Green Revolution period. During the post-Green Revolution period, foodgrain imports showed a decline, both in terms of quantity and value. The decline in quantity is at a much higher significant rate than the rate of decline in value.

On the whole, foodgrain imports of India, in terms of quantity, showed a decline from 0.71 million tonnes in 1955-56 to 0.13 million tonnes in 1986-87, *i.e.*, at a rate of 6.30 per cent per annum. During the same period, foodgrain imports in terms of value showed an increase from Rs. 35.09 crores to Rs. 47.00 crores, *i.e.*, at a rate of 2.02 per cent per annum, but not significant.

4.5. Pattern of Foodgrain Imports of India

An in-depth study of the pattern of imports of foodgrains of India enables us to understand the different types of foodgrains which are being imported and the relative importance of each of the foodgrains imported.

---

13. GOI., *Economic Survey 1982-83*, Delhi, p.6.
14. *Ibid.*, p.70
15. *Ibid.*, p.64
16. *Ibid.*, p.88

The relative shares of the different categories of foodgrains imported by India, in terms of quantity and value, in the total foodgrain imports of India are presented in Table 4.9, with a view to examine the relative importance of different categories of foodgrains in the total foodgrain imports and analyse the changes that have taken place in the pattern of foodgrain imports of India overtime.

From Table 4.9, it is evident that rice and wheat are the major foodgrains imported by India. Rice accounted for 37.27 per cent in terms of quantity and 13.39 per cent in terms of value, of the total foodgrains imported by India in 1955-56. The relative share of rice declined from 37.27 per cent in 1955-56 to 7.60 per cent in 1966-67, *i.e.*, during the pre-Green Revolution period. During the post-Green Revolution period, *i.e.*, from 1967-68 to 1986-87, the relative share of rice imports in the total foodgrain imports declined from 5.22 per cent to 2.33 per cent, with a peak level of 36.58 per cent in 1984-85. The relative share of rice imports, in terms of value, in the total imports of foodgrains increased from 13.39 per cent in 1955-56 to 15.60 per cent in 1966-67 during the pre-Green Revolution period. The relative share of rice imports in the total value of foodgrains imports declined from 10.29 per cent in 1967-68 to 2.13 per cent in 1986-87, during the post-Green Revolution period.

The relative share of imports of wheat in terms of quantity increased from 61.18 per cent in 1955-56 to 75.15 per cent in 1966-67, during the pre-Green Revolution period and further to 83.72 per cent in 1986-87 during the post-Green Revolution period. The relative share of wheat in terms of value increased from 34.43 per cent in 1955-56 to 80.87 per cent in 1966-67, during the pre-Green Revolution period, but declined to 78.72 per cent in 1986-87 during the post-Green Revolution period.

The relative share of other foodgrain imports in terms of quantity increased from 1.55 per cent to 17.25 per cent during the pre-Green Revolution period, 1955-56 to 1966-67, but the relative share in terms of value declined from 52.18 per cent in 1955-56 to 3.53 per cent in 1966-67. During the post-Green Revolution period, the relative share of the other foodgrain imports declined from 21.58 per cent in 1967-68 to 13.95 per cent in 1986-87 in terms of quantity but showed a marginal increase from 18.59 per cent to 19.15 per cent, in terms of value.

Table 4.9 : Pattern of Foodgrain Imports of India : 1955-56 to 1986-87
(Percentage Distribution in Terms of Quantity and Value)

Period	Rice		Wheat		Other Foodgrains		Total Foodgrains	
	Quantity	Value	Quantity	Value	Quantity	Value	Quantity	Value
(1)	(2)	(3)	(4)	(5)	(6)	(7)	(8)	(9)
1955-56	37.27	13.39	61.28	34.43	1.55	52.18	100.00	100.00
1956-57	22.95	55.56	77.05	43.27	—	—	100.00	100.00
1957-58	20.26	10.41	77.65	21.40	2.09	68.19	100.00	100.00
1958-59	12.32	36.54	84.24	32.90	3.44	30.56	100.00	100.00
1959-60	7.63	6.43	91.85	78.75	0.52	14.82	100.00	100.00
1960-61	13.61	10.66	85.38	79.42	1.01	8.92	100.00	100.00
1961-62	10.99	11.52	8.47	54.54	0.54	33.94	100.00	100.00
1962-63	10.72	15.08	87.88	65.10	1.40	19.82	100.00	100.00
1963-64	10.46	11.12	88.18	56.02	1.36	32.86	100.00	100.00
1964-65	9.62	15.06	83.90	83.33	6.48	1.60	100.00	100.00
1965-66	10.49	11.71	88.22	82.22	1.29	6.07	100.00	100.00
1966-67	7.60	15.60	75.15	80.87	17.25	3.53	100.00	100.00
1967-68	5.22	10.29	73.20	71.12	21.58	18.59	100.00	100.00

1968-69	7.80	17.07	83.73	77.09	8.47	5.84	100.00	100.00
1969-70	12.58	23.03	79.80	72.86	7.62	4.12	100.00	100.00
1970-71	9.94	14.00	80.50	81.35	9.56	4.65	100.00	100.00
1971-72	11.83	13.78	84.71	78.20	3.46	8.02	100.00	100.00
1972-73	16.75	13.26	62.25	59.60	21.00	27.14	100.00	100.00
1973-74	0.82	1.36	69.77	73.14	29.41	25.50	100.00	100.00
1974-75	0.98	1.59	89.29	91.41	9.73	7.00	100.00	100.00
1975-76	2.62	2.72	93.82	89.90	3.56	7.38	100.00	100.00
1976-77	3.13	5.30	76.32	93.00	20.55	1.60	100.00	100.00
1977-78	4.17	34.12	85.94	60.07	9.89	5.81	100.00	100.00
1978-79	1.28	11.74	97.44	76.03	1.28	12.23	100.00	100.00
1979-80	0.98	7.02	81.95	79.69	17.07	13.29	100.00	100.00
1980-81	4.09	30.79	65.08	56.36	30.93	12.85	100.00	100.00
1981-82	4.27	4.24	87.20	94.25	8.53	4.68	100.00	100.00
1982-83	1.42	1.07	96.94	95.42	1.64	3.59	100.00	100.00
1983-84	7.71	12.36	72.15	79.48	18.14	8.16	100.00	100.00
1984-85	36.58	37.19	58.07	53.72	5.35	9.09	100.00	100.00
1985-86	14.81	12.73	59.26	55.45	25.93	31.82	100.00	100.00
1986-87	2.33	2.13	83.72	78.72	13.95	19.15	100.00	100.00

Source : Based on data from GOI., Bulletin on Food Statistics, Delhi; RBI., Report on Currency and Finance, various issues

On the whole, the relative share of rice imports during the period 1955-56 to 1986-87 declined both in terms of quantity and value while the relative share of wheat imports showed an increase. Imports of rice accounted for 2.33 per cent and imports of wheat accounted for 83.72 per cent and other foodgrains accounted for 13.95 per cent in 1986-87, in terms of quantity of the total foodgrains imported by India. The relative share of other foodgrains in the total foodgrain imports in terms of quantity increased from 1.55 per cent in 1955-56 to 13.95 per cent in 1986-87 while the relative share of other foodgrains in terms of value declined from 52.18 per cent in 1955-56 to 19.15 per cent in 1986-87. The relative share of other foodgrains was not very significant, during the period under consideration, except in some of the years. The relative shares of rice and wheat showed considerable fluctuations.

Trends in each of the major foodgrains imported by India, in terms of quantity and value are examined below.

4.5.1. *Imports of Rice*

Trends in the imports of rice in terms of quantity and value are shown in Columns (2) and (4) of Table 4.10. Three-Year moving averages are calculated for quantity and value are shown in Columns (3) and (5) of the same Table.

The coefficients of variation and the estimated exponential growth rates are shown in Table 4.11. It is evident from Tables 4.10 and 4.11, that imports of rice increased from 265 thousand tonnes in 1955-56 to 787 thousand tonnes in 1966-67, *i.e.*, at a rate of 7.03 per cent per annum, in terms of quantity during the pre-Green Revolution period.

India imported rice from a number of countries in 1967. However, the imports of rice from individual countries were not on a large scale. These imports formed only a small percentage of the internal production of rice.[17]

Rice imports showed a decline from 787 thousand tonnes in 1966-67 to 453 thousand tonnes in 1967-68. The reduced dependence

---

17. GOI., Economic Survey of Indian Agriculture 1966-67, Delhi, p.134.

Table 4.10: Trends in the Quantity and Value of Rice Imports of India : 1955-56 to 1986-87

Q - Thousand tonnes
V - Rs. in crores

Period	Quantity	3-year moving	Value average	3-year moving average
(1)	(2)	(3)	(4)	(5)
1955-56	265	—	4.70	—
1956-57	325	445	31.98	17.86
1957-58	745	489	16.90	30.70
1958-59	397	479	44.02	23.34
1959-60	295	464	9.11	25.19
1960-61	699	459	22.44	15.52
1961-62	384	491	15.02	19.57
1962-63	390	419	21.26	18.89
1963-64	483	506	20.40	27.25
1964-65	645	637	40.10	32.74
1965-66	783	738	37.71	53.15
1966-67	787	674	81.64	58.04
1967-68	453	562	54.76	64.62
1968-69	446	462	57.47	56.83
1969-70	487	431	58.25	48.51
1970-71	361	364	29.82	35.28
1971-72	243	246	18.07	19.53
1972-73	134	263	10.71	11.74
1973-74	36	75	6.45	9.78
1974-75	55	96	12.17	18.38
1975-76	198	152	36.52	31.56
1976-77	204	143	46.00	41.44
1977-78	27	78	41.80	32.67
1978-79	3	11	10.20	19.81
1979-80	4	8	7.43	16.18
1980-81	18	29	30.91	17.68
1981-82	65	13	14.69	16.53
1982-83	19	164	4.00	39.56
1983-84	409	259	100.00	64.33
1984-85	349	267	90.00	64.67
1985-86	44	132	14.00	35.00
1986-87	3	—	1.00	—

Source : 1. GOI., Bulletin on Food Statistics, various issues, Delhi.
2. GOI., Statistical Abstract: India, various issues.
3. UN., F.A.O., Trade Year Book, various issues, Rome.

Table 4.11 : Growth Rate of Quantity and Value of Rice Imports of India : 1955-56 to 1986-87

Period	Quantity		Value	
	Gr.	CV	Gr.	CV
1955-56 to 1966-67	7.03 ** (2.57)	38.91	13.02 ** (2.48)	71.91
1967-68 to 1986-87	-15.43 * (2.65)	99.68	-6.54 (1.51)	8.10
1955-56 to 1986-87	-11.09 * (4.88)	81.24	-1.47 (0.75)	82.34

Note: Gr. = Exponential growth rate.
CV = Coefficient of variation.
Figures in brackets represent 't' values.
* Significant at 1 per cent level.
** Significant at 5 per cent level.

on imports in 1967-68 was the result of improvement on the domestic food front and consequent fall in food imports. Rice imports of India declined substantially from 487 thousand tonnes in 1969-70 to 36 thousand tonnes in 1973-74, but increased again to 198 thousand tones in 1975-76. The record harvest of 1975-76 and the large buffer stocks made it relatively less essential to import more food in the years following 1976-77. Rice imports declined substantially from 204 thousand tonnes in 1976-77 to 4 thousand tones in 1979-80. Rice imports increased sharply to 409 thousand tonnes in 1983-84, but declined to 3 thousand tonnes in 1986-87. Thus, rice imports of India declined substantially from 453 thousand tonnes in 1967-68 to 3 thousand tonnes in 1986-87, *i.e.*, at a rate of 15.43 per cent per annum, during the post-Green Revolution period.

On the whole, rice imports of India declined from 265 thousand tonnes in 1955-56 to 3 thousand tonnes in 1986-87, *i.e.*, at a rate of 11.09 per cent per annum, with a peak level of 787 thousand tonnes in 1966-67.

Imports of rice in terms of value increased from Rs. 4.70 crores in 1955-56 to Rs. 81.64 crores in 1966-67, *i.e.*, at a rate of 13.02 per cent per annum during the pre-Green Revolution period. On the whole, the value of rice imports decreased from Rs. 4.70 crores in 1955-56 to Rs. 1.00 crore in 1986-87. The growth rate, however, is not found to be significant.

4.5.2. *Imports of Wheat*

Wheat constitutes the major item in the foodgrain import bill of India. Imports of wheat of India in terms of quantity and value are shown in Columns (2) and (4) of Table 4.12. Three-year moving averages are calculated for quantity and value of imports of wheat and shown in Columns (3) and (5) of the same Table. Coefficients of variation and the estimated exponential growth rates for the quantity and value of wheat imports are shown in Table 4.13. It is evident from Tables 4.12 and 4.13 that, wheat imports of India increased from 0.435 million tonnes in 1955-56 to 7.784 million tonnes in 1966-67, *i.e.*, at a rate of 18.88 per cent per annum, in the pre-Green Revolution period.

The most important supplier of wheat to India was USA., from whom India imported considerable quantities of wheat under PL 480 Agreement. The other important countries, which supplied wheat to India in 1967 were Australia and Canada. The unprecedented drought of 1965-66 and 1966-67 created shortages in domestic production and lead to food imports in a big way.[18]

The comfortable food situation in 1967-68 season paved the way for a substantial reduction in the imports of wheat in 1968. Wheat imports declined substantially from 6.348 million tonnes in 1967-68 to 4.766 million tonnes in 1968-69, 3.09 million tonnes in 1969-70, 2.923 million tonnes in 1970-71 and 1.740 million tonnes in 1971-72. However, wheat imports of India increased steeply from 0.498 million tonnes in 1972-73 to a high level of 7.094 million tonnes in 1975-76, but declined to 0.284 million tonnes in 1980-81.

Wheat imports increased again from 0.284 million tonnes in 1980-81 to a high level of 3.039 million tonnes in 1983-84. The drought of 1982 led to a sharp increase in wheat imports. The quantity of imports of wheat declined from 3.039 million tonnes in 1983-84 to 0.108 million tonnes in 1986-87, due to the comfortable stock position of foodgrains. Thus, wheat imports declined substantially from 6.348 million tonnes in 1967-68 to 0.108 million tonnes in 1986-87, *i.e.*, at a rate of 14.72 per cent per annum, in the post-Green Revolution period.

---

18. GOI., *Economic Survey of Indian Agriculture 1966-67*, 1968, Delhi, pp.133-134.

Table 4.12 : Trends in the Quantity and Value of Wheat Imports of India : 1955-56 to 1986-87

Period	Quantity	3-year moving average	Value	3-year moving average
(1)	(2)	(3)	(4)	(5)
1955-56	435	—	12.08	—
1956-57	1,091	1,461	24.36	23.73
1957.58	2,856	2,221	34.75	53.92
1958-59	2,716	3,042	39.65	82.92
1959-60	3,553	3,552	111.35	122.40
1960-61	4,385	3,677	153.20	111.70
1961-62	3,092	3,559	70.55	105.20
1962-63	3,199	3,455	91.86	88.42
1963-64	4,073	2,598	102.86	138.87
1964-65	5,621	5,426	221.90	196.50
1965-66	6,583	6,663	264.74	377.19
1966-67	7,784	6,905	423.04	355.42
1967-68	6,348	6,299	378.47	353.67
1968-69	4,766	4,735	259.49	274.09
1969-70	3,090	3,593	184.33	205.73
1970-71	2,923	2,584	173.37	153.43
1971-72	1,740	1,720	102.60	108.04
1972-73	498	3,261	48.16	165.62
1973-74	3,070	2,856	346.10	364.15
1974-75	5,000	5,055	698.18	750.50
1975-76	7,094	5,689	1207.22	904.17
1976-77	4,972	4,207	807.10	695.97
1977-78	556	1,919	73.58	315.97
1978-79	229	374	6.07	74.65
1979-80	336	283	84.31	68.99
1980-81	284	694	56.59	146.88
1981-82	1,328	971	299.75	236.03
1982-83	1,301	1,889	351.74	431.50
1983-84	3,039	1,631	643.00	374.91
1984-85	554	1,256	130.00	278.00
1985-86	176	279	61.00	76.00
1986-87	108	—	37.00	—

Source: 1. GOI., *Statistical Abstract: India*, various issues, Delhi.
2. GOI., *Bulletin on Food Statistics*, various issues, Delhi.
3. RBI., *Report on Currency and Finance*, various issues Bombay.

Table 4.13 : Growth Rates of Quantity and Value of Wheat Imports of India: 1955-56 to 1986-87

Period	Quantity		Value	
	Gr.	CV	Gr.	CV
1955-56 to 1966-67	18.88 * (5.25)	55.80	26.92 * (7.88)	93.75
1967-68 to 1986-87	-14.72 * (3.70)	94.35	-4.61 (1.15)	104.37
1955-56 to 1986-87	-6.53 * (3.18)	78.09	3.73 ** (1.84)	114.03

Note: Gr. = Exponential growth rate.
CV = Coefficient of variation.
Figures in brackets represent 't' values.
* Significant at 1 per cent level.
** Significant at 10 per cent level.

On the whole, imports of wheat declined substantially from 0.435 million tonnes in 1955-56 to 0.108 million tonnes in 1986-87, *i.e.*, at a rate of 6.53 per cent per annum. In terms of value wheat imports increased from Rs. 12.08 crores in 1955-56 to Rs. 37.00 crores in 1986-87, *i.e.*, at a rate of 3.73 per cent per annum. wheat imports showed considerable fluctuations both in terms of quantity and value during this period.

4.5.3. *Imports of Other Foodgrains of India*

Among other foodgrains, India imported certain amount of maize, other cereals unmilled, grams and other pulses. Data relating to imports of other foodgrains, in terms of quantity and value, for the period 1955-56 to 1986-87 are given in columns (2) and (4) of Table 4.14. Three-year moving averages of quantity and value are shown in Columns (3) and (5) of the same Table, for which exponential growth rates are estimated. The coefficients of variation and the estimated exponential growth rates are shown in Table 4.15.

It is evident from Tables 4.14 and 4.15 that the three year moving average of imports of other foodgrains increased from 0.69 million tonnes in 1958-59 to 1.251 million tonnes in 1966-67, *i.e.*, at a rate of 40.73 per cent per annum, in the pre-Green Revolution period.

Table 4.14 : Trends in Imports of Other Foodgrains of India : 1955-56 to 1986-87

Q - Thousand tonnes
V - Rs. in crores

Period	Quantity	3-year moving average	Value	3-year moving average
(1)	(2)	(3)	(4)	(5)
1955-56	11	—	18.31	—
1956-57	—	—	—	—
1957-58	77	—	110.74	—
1958-59	111	69.33	36.83	56.17
1959-60	20	61.00	20.95	24.99
1960-61	52	30.33	17.20	27.38
1961-62	19	40.66	43.99	70.51
1962-63	51	44.33	27.97	44.10
1963-64	63	182.77	60.34	30.85
1964-65	434	197.77	4.25	28.05
1965-66	96	772.33	19.55	14.08
1966-67	1,787	1,251.33	18.45	45.64
1967-68	1,871	1,380.00	98.93	45.68
1968-69	482	882.77	19.66	43.00
1969-70	295	374.77	10.42	13.30
1970-71	347	237.77	9.81	10.25
1971-72	71	195.33	10.53	14.09
1972-73	168	511.00	21.93	51.04
1973-74	1,294	669.00	120.65	65.36
1974-75	545	702.77	53.51	91.07
1975-76	269	717.77	99.06	55.49
1976-77	1,339	557.33	13.90	40.03
1977-78	64	468.77	7.12	10.56
1978-79	3	45.77	10.67	10.62
1979-80	70	68.77	14.06	12.54
1980-81	133	111.00	12.90	19.84
1981-82	130	95.00	32.56	20.97
1982-83	22	305.33	17.46	38.67
1983-84	764	279.00	66.00	35.15
1984-85	51	297.33	22.00	41.00
1985-86	77	48.77	35.00	22.00
1986-87	18	—	9.00	—

Source: 1. GOI., *Statistical Abstract : India*, various issues, Delhi.
2. GOI., *Bulletin on Food Statistics*, various issues, Delhi.
3. RBI., *Report on Currency and Finance*, various issues, Bombay.
4. U.N., FAO., *Trade Year Book*, various issues, Rome.
5. UN., *International Trade Statistics*, various issues, Delhi.

Table 4.15 : Growth Rates of Other Foodgrains Imports of India (3 -Year moving average)

Period	Quantity		Value	
	Gr.	CV	Gr.	CV
1958-59 to 1967-68	40.73 * (4.06)	145.48	-5.55 (0.87)	46.43
1967-68 to 1985-86	-10.84 * (3.17)	82.27	--0.86 (0.29)	65.49
1958-59 to 1986-87	1.74 (0.64)	97.72	-1.54 (1.04)	58.50

Note: Gr. - Exponential growth rate.
CV - Coefficient of variation.
Figures in brackets represent 't' values
* Significant at 1 per cent level.

Other foodgrain imports declined from 1.380 million tonnes in 1967-68 to 0.049 million tonnes in 1985-86, *i.e.*, at a rate of 10.84 per cent per annum, during the post-Green Revolution period. Growth rate is not found to be significant for imports of other foodgrains in terms of value.

4.6 Direction of Imports of Foodgrains of India

An attempt is made here to analyse the major sources of foodgrain imports of India in order to identify the principal countries from which foodgrains are imported. The percentage share of each country in the total imports of each of the foodgrain imports of considered and the countries are ranked on the basis of these shares. The quantity imported and the percentage share of the first two countries, with the remaining countries grouped under the category of 'others' are presented in Table 4.16. The analysis is presented for the major foodgrains imported by India.

4.6.1. *Direction of Rice Imports*

It is evident from Table 4.16 that in 1955-56, India imported 265 thousand tonnes of rice entirely from Burma. In 1960-61, Burma continued to be the major supplier of rice to India. Rice imports from Burma increased from 265 thousand tonnes in 1955-56 to 336 thousand tonnes in 1960-61. Burma accounted for 48.07 per cent of the total rice imports of 699 thousand tonnes in 1960-61, followed by

Table 4.16 : Direction of Imports of Foodgrains of India

(Quantity: Thousand tonnes)

Commodity (1)	Origin of Imports (2)	Quantity imported (3)	Percentage share (4)
		1955-56	
Rice	Burma	265	100.00
	Total	265	100.00
Wheat and Wheat flour	Australia	306	70.34
	Others	129	29.66
	Total	435	100.00
Milo	U.S.A.	11	100.00
	Total	11	100.00
		1960-61	
Rice	Burma	336	48.07
	U.S.A	257	36.77
	Others	106	15.16
	Total	699	100.00
Wheat	U.S.A	4,040	92.11
	Australia	322	7.34
	Others	23	0.55
	Total	4,385	100.00
Milo	U.S.A.	52	100.00
	Total	52	100.00
		1965-66	
Rice	U.S.A.	286	36.53
	Burma	204	26.05
	Others	293	37.42
	Total	783	100.00
Wheat	U.S.A	6,021	91.46
	Australia	324	4.92
	Others	238	3.62
	Total	6,583	100.00

(Contd.)

(1)	(2)	(3)	(4)
Milo and Sorghum	U.S.A.	96	100.00
	Total	96	100.00
	1971-72		
Rice	Burma	110	45.27
	Thailand	84	34.57
	Others	49	20.16
	Total	243	100.00
Wheat	U.S.A.	1,200	68.97
	Others	540	31.03
	Total	1,740	100.00
Other foodgrains	U.S.A	71	100.00
	Total	71	100.00
	1975-76		
Rice	Thailand	198	100.00
	Total	198	100.00
Wheat	U.S.A	4,631	62.28
	Australia	484*	6.82
	Others	1,979	27.90
	Total	7,094	100.00
Milo and others	Argentina	180	66.91
	Others	89	33.09
	Total	269	100.00
	1980-81		
Rice	Nepal	14,926	82.93
	Bangladesh	2,780	15.44
	Others	294	1.63
	Total	18,000	100.00
Wheat and meslin and seed	U.S.A	217	76.41
	Others	67	23.59
	Total	284	100.00

(*Contd.*)

(1)	(2)		(3)	(4)
Pulses	Nepal		85	63.91
	Others		48	36.09
	Total		133	100.00
		1985-86		
Rice	Thailand		43.7	99.32
	Others		0.3	0.68
	Total		44.0	100.00
Wheat and meslin and seed	U.S.A		176	100.00
	Total		176	100.00
Pulses	Burma		56	72.73
	Others		21.	27.27
	Total		77	100.00

Notes: * Includes Cereals received as gift.
** received as gift.

Source: 1. GOI., Bulletin on Food Statistics, various issues, Delhi.
2. RBI., Report on Currency and Finance, various issues, Bombay
3. U.N. FAO., Trade Year Book, various issues, Rome.

USA with 36.77 per cent, and UAR with 15.16 per cent (U.A.R being the only country under the group 'others'. In 1965-66, USA accounted for 36.53 per cent, followed by Burma with 26.05 per cent of the total rice imports of India of 783 thousand tonnes. Among the other countries, Thailand is the major supplier of rice to India.

In the year 1971-72, Burma accounted for 45.27 per cent and Thailand for 34.57 per cent and others for 20.16 per cent if the total rice imports of India of 243 thousand tonnes. In 1975-76, there were no imports of rice to India from Burma. Thailand supplied the entire 198 thousand tonnes of rice imported by India in this year.

In 1980-81, Nepal accounted for 82.93 per cent and Bangladesh accounted for 15.44 per cent and others accounted for 1.63 per cent of the total rice imports of India of 18,000 tonnes.

In 1985-86, Thailand accounted for 99.32 per cent others accounted for 0.68 per cent of the total rice imports of India of 44 thousand tonnes.

Thus, the rice imports of India were met mainly by Burma, Thailand, UAR and USA.

4.6.2 *Direction of Wheat Imports*

Direction of wheat imports of India are shown in Table 4.16. It is evident from the Table that Australia accounted for 70.34 per cent and others accounted for 29.66 per cent of the total wheat imports of 435 thousand tonnes of India in 1955-56. In 1960-61, USA accounted for 92.11 per cent, Australia accounted for 7.34 per cent and others accounted for 0.55 per cent of the total wheat imports of 4,386 thousand tonnes to India. In the year 1965-66, USA accounted for 91.46 per cent under PL 480 Agreement, Australia accounted for 4.92 per cent and others accounted for 3.62 per cent of total wheat imports of 6,583 thousand tonnes of India.

In the year 1971-72, USA accounted for 68.97 per cent and others accounted for 31.03 per cent of total wheat imports of 1,740 thousand tonnes of India. In 1975-76, USA accounted for 62.28 per cent, while Australia accounted for 6.82 per cent and others accounted for 27.90 per cent of the total wheat imports of 7,094 thousand tonnes of India.

In 1980-81, USA accounted for 76.41 per cent and others accounted for 23.59 per cent of the total wheat imports of 284 thousand tonnes of India. In 1985-86, the total imports of 176 thousand tonnes of India were met by U.S.A.

Thus, USA, Australia and Canada are the major suppliers of wheat imports of India during the period 1955-56 to 1985-86.

4.6.3 *Direction of Other Foodgrain Imports*

Apart from wheat and rice, India imported other foodgrains such as milo or sorghum, maize, other cereals unmilled, grams and other pulses, from abroad. During 1950s, 1960s and 1970s, milo was the major foodgrain of other foodgrain imports. India imported milo from USA, Australia, China, Argentina and Pakistan. Imports of milo and sorghum increased from 11 thousand tonnes in 1955-56 to 269 thousand tonnes in 1975-76.

During 1980s, there is a change in pattern and direction of imports of other foodgrains. The imports of milo declined during the eighties, while the imports of other pulses and gram increased significantly due to the domestic shortfalls in the production of pulses and gram.

In 1980-81, Nepal accounted for 63.91 per cent, and others accounted for 36.09 per cent of the total pulses imports of 133 thousand tonnes of India. In the year 1985-86, Burma accounted for 72.73 per cent while other countries accounted for 27.27 per cent of the total pulses and gram imports of 77 thousand tonnes of India.

Thus, the USA was the sole supplier of milo to India, followed by Argentina and Australia upto 1976-77. It is evident from Table 4.16 that the increased imports of milo played a role to some extent in reducing the imports of rice and wheat which were relatively costly. It is interesting to note that there is a change in the pattern and direction of imports of other foodgrains from 1975-76 onwards. It indicates that there is an increase in pulses and other foodgrain imports from other countries, in the place of imports of rice and wheat.

4.7. Import Dependence of Foodgrain Imports of India

Before Independence, shortage of foodgrains has been recognised as one of the important problems faced by the Indian economy. After the partition of India, the problem was highlighted due to various unforeseen difficulties such as low acreage of food crops, increase in population and high income levels. The consumption of foodgrains both for household purposes and for industrial purposes has been increasing, while the domestic production of foodgrains is relatively low due to drought and flood conditions. The gap between these two, *i.e.*, supply and demand, is widening over time, necessitating imports on a large scale. Several measures and policies have been taken by the Government from time to time to reduce the imports of foodgrains and to increase the domestic production of foodgrains. The import-availability ratio(M/S) of foodgrains indicates the progress of import substitution that has taken place overtime as a result of the measures taken by the Government, where:

M = imports of foodgrains
P = domestic production of foodgrains
S = M + P = total availability of foodgrains,

The data relating to production, imports, availability and the ratio of import-availability for total foodgrains during the period 1955-56 to 1986-87 are presented in Table 4.17. From the Table, it is evident that the average production of foodgrains increased from 72.84

Table 4.17 : Import Dependence of India on Foodgrains Imports

(Million tonnes)

S.No.	Contents	1955-56 to 1959-60	1960-61 to 1964-65	1965-66 to 1969-70	1970-71 to 1974-75	1975-76 to 1979-80	1980-81 to 1984-85	1985-86 to 1986-87
1.	Production of Foodgrains (P)	72.84	83.04	87.02	103.00	120.04	138.06	146.90
2.	Total Availability (S) of Foodgrains	2.58	4.72	7.21	3.30	3.07	1.69	0.21
3.	Import availability ratio (M/S)	0.03	0.05	0.08	0.03	0.02	0.01	0.001

Source: Based on GOI., *Economic Survey*, Delhi, various issues.

million tonnes during 1955-56 to 146.90 million tonnes during 1985-87, (the latest period for which comparable data are available), while the average imports of foodgrains decreased from 2.58 million tonnes to 0.21 million tonnes during this period. The average import-availability ratio (M/S) of total foodgrains increased from 0.03 during 1955-56 to 0.05 during 1960-65 and further to 0.08 during 1965-70, but declined to 0.03 during 1970-75. It decreased to a level of 0.01 during 1980-81 and further to 0.0001 during 1985-87, due to growth in the production of foodgrains.

On the whole, the M/S ratio for total foodgrains declined from 0.03 during 1955-60 to 0.0001 during 1985-87. The above analysis indicates that imports of foodgrains accounted for a small proportion of the total supplies. There is a reduction in the degree of import dependence of India on foodgrain imports during the period under study, *i.e.*, 1955-56 to 1986-87.

4.8 Summary

The analysis of Food and non-food imports shows that the relative share of Food imports is showing a decline and the relative share of non-Food items is also showing a decline, while, the relative shares of Manufactured Fertilisers and Petroleum Products are showing increase in the total imports of India during the period, 1955-56 to 1986-87.

Imports of foodgrains (or Cereals) come under the category of imports of Food and Live Animals Chiefly for Food. In 1985-86, imports of Cereals accounted for 12.89 per cent of imports of Food and Live Animals Chiefly for Food, and 0.56 per cent of the total imports of India.

Imports of Cereals and Cereal preparations increased from Rs. 35.09 crores in 1955-56 to Rs. 523.10 crores in 1966-67, *i.e.*, at a rate of 17.69 per cent per annum, during the pre-Green Revolution period. Cereals became the largest single item of the total imports of India in 1975-76 and accounted for more than a quarter of the total value of imports of India, exceeding the relative share of even the value of the imports of POL. However, imports of Cereals and Cereal preparations declined from Rs. 532.20 crores in 1967-68 to Rs. 47.00 crores in 1986-87, *i.e.*, at a rate of 4.12 per cent per annum, though not significant, during the post-Green Revolution period. On the whole, imports of cereals, in terms of value, increased from

Rs. 35.09 crores in 1955-56 to Rs. 47.00 crores in 1986-87, *i.e.*, at a rate of 2.02 per cent per annum, but not significant. During this period, imports of Food and Live Animals increased at a rate of 6.20 per cent per annum and total imports of India increased at a rate of 11.30 per cent per annum.

In terms of quantity, foodgrain imports of India increased from 0.71 million tonnes in 1955-56 to 10.36 million tonnes in 1966-67, *i.e.*, at a rate of 17.43 per cent per annum, during the pre-Green Revolution period. Foodgrain imports declined from 8.67 million tonnes in 1967-68 to 0.13 million tonnes in 1986-87, *i.e.*, at a rate of 14.30 per cent per annum, during the post-Green Revolution period. On the whole, foodgrain imports in terms of quantity, declined from 0.71 million tonnes in 1955-56 to 0.13 million tonnes in 1986-87, *i.e.*, at a rate of 6.30 per cent per annum.

Rice and wheat are the major foodgrains imported by India. Together, rice and wheat accounted for 86.05 per cent in terms of quantity and 80.85 per cent in terms of value of the total foodgrains imported by India in 1986-87. There is a decline in the relative share of rice and an increase in the relative share of wheat in the total foodgrains imported by India during the period, 1955-56 to 1986-87, indicating a change in the pattern of foodgrain imports of India and a shift towards greater wheat imports.

Rice imports of India, in terms of quantity declined at a rate of 11.90 per cent per annum while, wheat imports of India, in terms of quantity, declined at a rate of 6.53 per cent per annum during the period 1955-56 to 1986-87. Imports of both rice and wheat showed considerable fluctuations.

Analysis of direction of imports of foodgrains indicates that rice is mainly imported from Burma. Thailand, UAR and USA. The major suppliers of wheat to India are USA, Australia and Canada. Among the other foodgrains milo is supplied to India mainly by USA, Argentina and Australia. It is also interesting to note that there is an increase in the imports of pulses among other foodgrains.

The average import-availability (M/S) ratio for total foodgrains showed a decline from 0.03 during the period 1955-56 to 0.0001 in 1985-87. This indicates that imports of foodgrains accounted for a small portion of the total supplies of foodgrains, and that there is a reduction in the degree of import dependence of India on foodgrain imports during the period under study *i.e.*, 1955-56 to 1986-87.

CHAPTER 5

DETERMINANTS OF IMPORTS OF FOODGRAINS OF INDIA

5.1. Introduction

STAGNATION in agricultural production and shortage of foreign exchange resources are, in general, two of the important economic problems faced by the Indian economy. Persistent difficulties in respect of balance of payments force developing economies like India to tighten import controls and to use foreign exchange regulations. In this context, import management becomes important not only to improve the balance of payments position but also to achieve the overall planning targets in India.[1] Against this background, imports of foodgrains have to be examined in order to assess the impact of various factors on the imports of foodgrains as well as to provide broad guidelines for policy making. Hence, an attempt is made in this Chapter to estimate the import demand function for foodgrains and to examine the factors determining the demand for imports of foodgrains of India.

5.2. Analysis of Import Demand Functions

There are two approaches in the analysis of import demand functions. The first is the standard import demand approach. It considers the import demand for any commodity as a function of the level of income and the ratio of the import price to the domestic price of the commodity.

The second approach is known as the balance of payments approach. It establishes a behavioural relationship between the imports of the country and its foreign exchange receipts, as a part of

1. S.K. Patel, *Quarterly Econometric Analysis of Imports in India*, Hindustan Publishing Corporation, Delhi, 1981, p.21.

the balance of payments adjustment. This approach was mainly developed by Hemphill.[2]

5.2.1. *The Standard Import Demand Function* :

The Standard import demand function is similar to an ordinary demand function. It takes the following form:

$Q_{mi} = f(Y_i, P_m/P_d)$, where

Q_{mi} = quantity of imports of country *i*,

Y_i = real income of country *i* and

P_m/P_d = ratio of import price to domestic price of the commodity under consideration.

Thus, the demand for imports of country '*i*' during time '*t*' depends on the level of real national income and the ratio of import price to domestic price. The relationship of these variables is explained below.

Income

In general the quantity of imports of the level of real income are related positively. An increase in real income generally leads to an increase in the quantity demanded and a decline in real income generally leads to a decline in the demand for imports. Hence, the sign of the coefficient real income is expected to be positive.

Relative Price

The second important variable considered in the standard import demand function is the relative price or the ratio of import price to domestic price. The use of this variable assumes that there some degree of substitutability between imports and domestic goods. However, it should be noted that such substitutability between imported goods and domestic goods may not always be perfect. Import controls and import licensing, in many of the developing countries like India, are used to ban the importation of some of the goods or to restrict the quantum of the commodity that may be imported, in order to save the limited foreign exchange to facilitate the importation of other essential commodities. Quality differences

---

2. W.L. Hemphill, "The Effects of Foreign Exchange Receipts on Imports of Less Developed Countries", *I.M.F. Staff Papers*, November 1974, pp. 637-677,

also make the substitution between imported goods and domestic goods to be not relevant.

Empirical studies have made use of index of unit value for price, in the absence of actual prices or price offers of the imported goods. Relative price is measured as the ratio of per unit value of imports to the wholesale price of corresponding domestic goods. The theoretically expected sign of the coefficient of the relative price variable is negative. A rise in import price relative to domestic price will reduce the demand for imports and fall in import price relative to domestic price will increase the demand for imports.

Quantitative Restrictions

In addition to the above two variables, *viz.*, income and relative price, considered in the standard import demand function, it is also necessary to include in the import demand function an additional variable to take account of the effect of quantitative restrictions on imports. This is so because many of the developing countries impose a number of quantitative restrictions on their imports, for various reasons, as pointed out earlier. However, there are a number of difficulties that are faced in practice in quantifying the role played by these controls in determining the quantity of imports demanded. Different economists have used different variables as measures of quantitative restrictions.

Some of the variables used by economists[3] as proxies for quantitative restrictions in estimating the import demand functions are: level of foreign exchange reserves, export earnings, foreign aid flows and overseas assets. The assumption that is involved here is that import controls very inversely with the country's capacity to import. When the capacity to import rises, controls will be relaxed and imports will rise and when the capacity to import declines, controls will be tightened and imports will decline.

---

3. (a) N. Islam, "Experiments in Econometric Analysis of An Import Demand Function", Pakistan *Economic Journal*, September 1961.
   (b) M. Dutta, "A Prototype Model of India's Foreign Sector", *International Economic Review*, 1964.
   (c) Alejandro Diaz, *Exchange Rate Devaluation in a Semi-Industrialised Country: The Experience of Argentina 1955-61*, Cambridge, Massachusetts, 1965.

5.2.2. *Import Demand Function Studies on India*

In the studies on import demand functions for India, different economists have employed different variables. Moorty and Sastry[4] have taken index of industrial activity and prices as the independent variables. Narsimham's[5] study used disposable income as an index of domestic income, but has not considered price as a variable. Benerji[6] used prices and income as explanatory variables in explaining the variations in the quantum of imports. Dutta[7] employed index of industrial output and prices as explanatory variables. Marwah[8] used relative price and industrial economic activity as explanatory variables to estimate import demand functions. Wadhwa[9] considered prices and national income to explain the variations in the quantum of imports of India. Jayaraman[10] used relative prices, gross national product and net foreign assets as explanatory variables in the study of import demand functions. Majority of these studies estimated demand functions for imports in general, based on Standard International Trade Classification (SITC) as a part of the economy-wide econometric models.

Patel's study[11] apart from taking traditional variables, also tries to quantify the effects of quantitative restrictions on imports. Based on quarterly data, the import demand function as specified in the study relates the quantity of imports demanded by the country to the ratio of domestic price of import price, an appropriate real income or output variable and some measure of quantitative restrictions.[12]

---

4. V.N. Moorty and V.K. Sastry, "Elasticities of Demand For Certain Indian Imports and Exports", *Sankhya*, Vol. XI, December 1951, pp. 309-336.
5. N.V.A. Narsimham, *A Short-Term Planning Model For India*, Amsterdam, 1956.
6. H. Benerji, "Analysis of Import Demand", *Artha Vijnana*, Vol. 1, No. 3, September 1959, pp.259-69
7. M. Dutta, "Import Structure of India", *Review of Economics and Statistics*, Vol. XLVII, 1965, pp. 295-300.
8. K. Marwah, "An Econometric Model of India: Estimating Prices, Their Role and Sources of Change", *Indian Economic Review*, Vol. VII, No. 1, April 1972, pp. 53-91
9. C.D. Wadhwa, "Elasticities of Demand for Indian Exports and Imports and the Question of the Devaluation of the Indian Rupee," *Foreign Trade Review*, Jan-March 1974, Vol. VIII, No. 4, pp. 364-77.
10. T.K. Jayaraman, "Import Demand Function of India", *Foreign Trade Review*, Vol. XI, No. 4, Jan-March 1977, pp. 367-72
11. S.K. Patel, *Quarterly Econometric Analysis of Imports in India*, *op. cit.*
12. *Ibid*.

M. Aggarwal[13] used real gross domestic product, ratio of import price index to wholesale price index, total foreign exchange reserves and annual exchange rate variations of the major currencies in the study of import demand functions, for thirteen selected developing countries including India.

5.2.3. *Import Demand Function for Food Imports of India*

T.K Jayaraman[14] estimated import demand function for Food and Live Animals corresponding to SITC Section 'O' by employing the variables *viz.*, import quantum index number of Food and Live Animals, price of imports, domestic wholesale price index (PW), Gross National Product (GNP) and net foreign assets (NFA). In estimating the import function in double logarithmic form, the variables GNP/PW and NFA/PW have been lagged by one period. In addition to the above set of explanatory variables, a dummy variable in order to capture the influence of the devaluation of Indian rupee that took place in 1966 and a variable 'T' representing time trend have been added. However, Jayaraman's analysis relates to the period 1961-72 only.

P.C. Verma[15] estimated import function for Food imports of India by considering value of food imports as a function of per capita food production index and food stock changes over previous year. Food imports considered by Verma consist of food category itself, beverages and tobacco, vegetable and animals oils including oilseeds, Further, the study covers the decade 1960-61 to 1969-70 only.

Basudeb Biswas and Rati Ram[16] maintained that India's foodgrain imports are amenable to meaningful economic analysis, even though the imports are controlled by the Central Government. They specified the traditional import demand function and modified it by a simple governmental intervention instrument. They study

---

13. M. Aggarwal, "Devaluation, Determinants of International Trade Flows and Payments Imbalance", *The Indian Economic Journal*, Vol. 31, No. 3, Jan-March, 1984, pp. 24-33.
14. T.K. Jayaraman, *op. cit.*
15. P.C. Verma, "Import Structure of India, Bangladesh and Pakistan", *Foreign Trade Review*, Vol. XII, No. 4, Jan-March 1978, pp. 498-510.
16. B Biswas and Rati Ram, "Demand Function For India's Foodgrain Imports: some Elasticity Estimates". *The Indian Economic Journal*, Vol. 27, No.4, April-June 1980, pp.12-19.

considers quantity of foodgrains imported as a function of gross national product, price of imported foodgrains, wholesale prices of foodgrains, domestic production of foodgrains, opening level of central stocks of foodgrains and foreign exchange reserves The authors have also included a dummy variable taking the value one for the buffer stock years 1973, 1974 and 1975 and zero for earlier years since, the building of sizeable buffer stocks has become more important in the latter years. There study relates to the period 1950-51 to 1974-45.

With respect to imports of foodgrains, S.K. Patel[17] considered income, population, price of foodgrains, agricultural production index and stocks of foodgrains as explanatory variables in estimating the demand for imports of foodgrains of India. He considered imports of Food to be basically 'stress' imports, as India has failed historically to meet the domestic demand for food from internal sources. Hence, he maintained that foreign exchange constraint was unimportant for Food imports. The pressure of domestic demand was expressed through a variable of foodgrain prices. Real income and population variables were added to show the income and size effects. On the supply side, food production was taken as a major determinant for Food imports.

Public Distribution System (PDS) in foodgrains has become an important feature in Indian food management, after the severe drought of 1965-66 and 1966-67. Under such circumstances, Patel maintained that governmental procurement of food and creation of buffer stocks of foodgrains assumed importance. He, therefore, included food stock at the beginning of the time period as an explanatory variable in the Food import demand function. However, Patel's analysis relates only to the period 1959 to 1973 based on quarterly data. He has not included external assistance as one of the variables.

Gulati[18] estimated wheat import demand function for India by considering the following variables, *viz.*, imports of wheat in the year '*t*' in thousand tonnes, closing stock of all foodgrains with the Government as on 31st December of the previous calendar year in

---

17. S.K. Patel, *Quarterly Econometric Analysis of Imports of India*, *op.cit.*, pp. 55-58.
18. A. Gulati, *Agricultural Price Policy in India: An Econometric Approach (with Special Reference to Wheat)*, Concept publishing Company, New Delhi, 1987, pp. 110-112.

thousand tonnes, output of wheat in year '*t*' in thousand tonnes, ratio between landed price of imported wheat and unit value index of all imports with 1970-71 as the base year and external aid utilised during the financial year '*t*' in Rs. 10 million.

Gulati maintained that food import is generally assigned top priority, and therefore, the foreign exchange constraint on foodgrain imports does not operate in a way as it does for non-essential imports. Gulati's analysis relates to the period 1965-66 to 1979-80 and considers wheat imports only.

Ramesh Babu[19] specified import demand function for imports of Food and Live Animals, corresponding to Revised International Trade Classification (RITC) Section '1'. The variables considered are value of imports of Food and Live Animals at constant prices, relative price (price of imports to domestic price), gross domestic product of India, external assistance including PL 480 and index of food production. His analysis relates to the period 1960-61 to 1976-77 only. He has not considered foodgrain imports separately. Further, he has not considered population and stock of foodgrains as independent variables in the import demand function.

5.3. Specification of the Model of the Present Study

The studies reviewed in the previous section indicate that the demand for imports of foodgrains of India is influenced by the following factors *viz.*, national income, relative price *i.e.*, the ratio of import price to domestic price, quantitative restrictions, domestic production of foodgrains, stocks of foodgrains, population, external assistance including PL 480 and foreign exchange reserves. To explain the demand for foodgrains, the variables considered in the present study are: per capita income, relative price, domestic production of foodgrains, stocks of foodgrains lagged by one year. Procurement of foodgrains, external assistance and net foreign assets. In mathematical notation, the import demand function for foodgrains, then, can be written as follows:

$$Q_{mft} = f(PY_t, RP_{ft}, P_{ft}, S_{f(t-1)}, PR_{ft}, EA_t, NFA_t) \qquad ...5.1$$

19. S. Ramesh Babu, *India's Foreign Trade: Some Trends*, Chugh Publications, Allahabad, 1988, pp.130-132.

Where,

Q_{mft} = Index number of quantum of imports of foodgrains of India in year '*t*',

PY_t = Index of per capita income in year '*t*',

RP_{ft} = Index of ratio of import price of foodgrains to domestic wholesale price of foodgrains in year '*t*',

P_{ft} = Index of domestic production of foodgrains in year '*t*',

$S_{f(t-1)}$ = Index of closing stocks of foodgrains with Government (Centre and States) in year '*t*-1',

PR_{ft} = Index of procurement of Foodgrains in year '*t*',

EA_t = Index of external assistance in year '*t*' and

NFA_t = Index of net foreign assets in year '*t*'.

5.3.1. *Measurement of the Variables and the Expected Signs of the Coefficients*

Variables considered in the present study are examined below.

Import of foodgrains is considered in terms of quantity in order to eliminate the price effect, which might arise when expressed in value terms. Import of foodgrains for the period 1955-56 to 1986-87 is the expressed in index from with base 1970-71 = 100, and is represented symbolically as 'Q_{mf}'. This is, then, taken to represent the dependent variable.

The first independent variable considered is per capita income. This is taken to represent income or economic activity which is the first important factor determining the demand for imports. Income is expressed in per capita terms to take into account the effect of population. The expected sign of the coefficient of the variable *viz.*, per capita income is positive. This implies that an increase in domestic income or economic activity increases the demand for imports of foodgrains and *vice versa* Per capita income in 1970-71 prices is obtained for the period 1955-56 to 1986-87 and expressed in an index form with 1970-71 = 100. It is denoted symbolically as '*PY*'.

The second independent variable considered in the present study is the relative price, *i.e.*, the ratio of import price to domestic price.

This is an important variable which determines the demand for imports, even in economies like India, which rely on import controls to a considerable extent. The theoretically expected sign of the coefficient of *Pm/Pd* is negative, indicating substitutability between imports and domestic production. Empirical studies have made use of unit values to represent prices. The per unit import price of foodgrains and domestic wholesale price of foodgrains are used to compute the relative price variable. This is then expressed in an index from with 1970-71 as the base year and is denoted as 'RP_f'.

The third variable considered to estimate the demand function for imports of foodgrains is domestic production of foodgrains, which influences the import of foodgrains to a significant extent. The expected sign of the coefficient of this variable is negative indicating that an increase in the production of foodgrains tends to reduce imports of foodgrains and *vice-versa*. Production of foodgrains is expressed in an index form with the triennium ending 1969-70 as the base year and is denoted as 'P_f'.

The stock of foodgrains plays an important role in the demand for imports of foodgrains. This is so because of the policy of the Government to supply foodgrains at subsidised prices through the Public Distribution System (PDS) and/or to maintain buffer stocks for food security. When the stocks of foodgrains with the Government fall below a certain level, the Government has to resort to imports of foodgrains. Therefore, we have included the stocks of foodgrains as an explanatory variable in the import demand function for foodgrains. It is assumed here that the imports of foodgrains are influenced by the level of closing stocks of foodgrains with the Government in the previous period. The closing stocks of foodgrains with the Government in million tonnes is expressed in index form with 1970-71 as the base year and lagged by one year, it is denoted as '$S_{f(t-i)}$'. The expected sign of the coefficient of the stock variable is negative indicating that as the closing stocks of foodgrains with the Government in the previous year are at a high level, the need to import foodgrains in the current years tends to decline and *vice-versa*.

Another variable considered in the import demand function for foodgrains in the present study is the procurement of foodgrains. Procurement of foodgrains gradually substitutes for imports of foodgrains. As the procurement of foodgrains increases, import of foodgrains tends to decline. Hence, the expected sign of the coefficient of procurement variable in negative.[20] Procurement of foodgrains in million tonnes is expressed in an index form with 1970-71 as the base year and is donoted by 'PR_t'.

The availability of external assistance including the tied-aid under PL 480 agreement is an important factor determining the imports of foodgrains of India. Therefore, external assistance[21] is considered as an important variable in the import demand function. External assistance, including PL 480, in value terms, is expressed in an index from and its denoted as 'EA_t'. The expected sign of the coefficient of this variable is positive, indicating that an increase in external assistance tends to increase the imports of foodgrains.

The last explanatory variable considered in the present study is net foreing assets. It is generally found that imports of developing countries like India are considerably influenced by import controls. Conventional variables may not be able to explain fully the variations in the level of imports. It is, therefore, necessary to introduce a suitable variable which reflects the import control measures. Net foreign assets, which includes foreign exchange reserves, gold and Special Drawing Rights (SDRs) of the country, is generally used as a proxy variable for import controls. A favourable net foreign assets position is generally associated with a relaxation of import controls and a depletion of net foreign assets is generally associated with tightening of import controls. Net foreign assets of India, in value terms, is expressed in an index from with 1970-71 as the base year and denoted as 'NFA'. The expected sign of the coefficient of the variable 'NFA' is negative.

The building up of buffer stocks in sizeable quantities has become relatively more important from the years 1973, 1974 and 1975. To study the shift in the import demand function, as a result of

---

20. However, when the procurement of foodgrains within the country is not at the required level, Government may decide to import foodgrains. In such a case, the coefficient of procurement variable may be positive, indicating increase in imports of foodgrains along with an increase in procrument. The use of lag in respect of procurement variable has not yielded good results and hence not employed here.
21. It may be noted here that external assistance represents the total among that could be used for the importation of all commodities and not foodgrains alone, as data could not be obtained separately for foodgrains.

policy change, a dummy variable 'D' is considered taking the value one of the year 1975-76[22] and zero for the earlier years has been introduced. Further to capture the effects of the policy changes relating to stocks and procurement of foodgrains on imports of foodgrains, slope dummy variables '$DS_{f(t-1)}$' and 'DP' have also been considered in the present study.

The analysis has been carried out for the period 1955-56 to 1986-87, as comparable data for all the variables considered are available only for this period. Correlation coefficients have been calculated for all the variables under consideration. The zero order correlation matrix[23] is given in Appendix Table 5.1.

5.3.2. *Alegebraic Specification of the Model*

Both linear and log-linear forms of the import demand function, which is specified in Section 5.3, have been tried with various combinations of the independent variables. The functions estimated can be written algebraically as follows:

$$Q_{mft} = a + b_1 PY_t \qquad .....(5.2)$$

$$\text{Log } Q_{mft} = a + b_1 \log PY_t \qquad .....(5.3)$$

$$Q_{mft} = a + b_1 RP_{ft} \qquad .....(5.4)$$

$$\text{Log } Q_{mft} = a + b_1 \log RP_{ft} \qquad ....(5.5)$$

$$Q_{mft} = a + b_1 PY_t + b_2 RP_{ft} \qquad ...(5.6)$$

$$\text{Log } Q_{mft} = a + b_1 \log PY_t + b_2 \log RP_{ft} \qquad ....(5.7)$$

$$Q_{mft} = a + b_1 RP_{ft} + b_2 P_{ft} \qquad ....(5.8)$$

$$\text{Log } Q_{mft} = a + b_1 \log RP_{ft} + b_2 \log P_{ft} \qquad ....(5.9)$$

$$Q_{mft} = a + b_1 RP_{ft} + b_2 S_{f(t-1)} \qquad ..(5.10)$$

22. The policy changes seem to have become effective from the year 1975 onwards. This is reflected in the data relating to stocks and procurement of foodgrains as given in Table 3.15 of Chapter III of the present study.

23. The zero order correlation matrix as given in Appendix Table 5.1, shows a high degree of correlation for some of the independent variables. The high correlation and the existence of the problem of multicollinearity have been pointed out by Biswas and Rati Ram in their study on foodgrain imports (See B. Biswas and Rati Ram, *op. cit.*, p 15) and S.K. Patel in his analysis of imports of India (See S.K. Patel, *op. cit.*, pp. 181-182). Efforts to overcome the problem of multicollinearity have not yielded meaningful results. This may, therefore, be considered as a statistical limitation of our study.

$$\text{Log } Q_{mft} = a + b_1 \log RP_{ft} + b_2 \log S_{f(t-1)} \quad ...(5.11)$$

$$Q_{mft} = a + b_1 P_{ft} + b_2 PR_{ft} \quad ...(5.12)$$

$$\text{Log } Q_{mft} = a + b_1 \log P_{ft} + b_2 \log PR_{ft} \quad ...(5.13)$$

$$Q_{mft} = a + b_1 PY_t + b_2 RP_{ft} + b_3 P_{ft} \quad ...(5.14)$$

$$\text{Log } Q_{mft} = a + b_1 \log PY_t + b_2 \log RP_{ft} + b_3 \log P_{ft} \quad ...(5.15)$$

$$Q_{mft} = a + b_1 RP_{ft} + b_2 P_{ft} + b_3 S_{f(t-1)} \quad ....(5.16)$$

$$\text{Log } Q_{mft} = a + b_1 \log RP_{ft} + b_2 \log P_{ft} + b_3 \log S_{f(t-1)} \quad ...(5.17)$$

$$Q_{mft} = a + b_1 P_{ft} + b_2 S_{f(t-1)} + b_3 PR_{ft} \quad ...(5.18)$$

$$\text{Log } Q_{mft} = a + b_1 \log P_{ft} + b_2 \log S_{f(t-1)} + b_3 \log PR_{ft} \quad ...(5.19)$$

$$Q_{mft} = a + b_1 P_{ft} + b_2 S_{f(t-1)} + b_3 EA_t \quad ...(5.20)$$

$$\text{Log } Q_{mft} = a + b_1 \log P_{ft} + b_2 \log S_{f(t-1)} + b_3 \log EA_t \quad ...(5.21)$$

$$Q_{mft} = a + b_1 RP_{ft} + b_2 S_{f(t-1)} + b_3 S_{f(t-1)} + b_4 EA_t \quad ...(5.22)$$

$$\text{Log } Q_{mft} = a + b_1 \log RP_{ft} + b_2 \log P_{ft} + b_3 \log S_{f(t-1)} + b_4 \log EA_t \quad ...(5.23)$$

$$Q_{mft} = a + b_1 RP_{ft} + b_2 P_{ft} + b_3 S_{f(t-1)} + D \quad ...(5.24)$$

$$\text{Log } Q_{mft} = a + b_1 \log RP_{ft} + b_2 \log P_{ft} + b_3 \log S_{f(t-1)} + D \quad ....(5.25)$$

$$Q_{mft} = a + b_1 RP_{ft} + b_2 P_{ft} + b_3 S_{f(t-1)} + D + DS_{f(t-1)} \quad ....(5.26)$$

$$\text{Log } Q_{mft} = a + b_1 \log RP_{ft} + b_2 \log P_{ft} + b_3 \log S_{f(t-1)} + D + DS_{f(t-1)} \quad ...(5.27)$$

$$Q_{mft} = a + b_1 RP_{ft} + b_2 P_{ft} + b_3 S_{f(t-1)} + b_4 EA_t + D + DS_{f(t-1)} \quad ....(5.28)$$

$$\text{Log } Q_{mft} = a + b_1 \log RP_{ft} + b_2 \log P_{ft} + b_3 \log S_{f(t-1)} + b_4 \log EA_t + D + DS_{f(t-1)} \quad .....(5.29)$$

$$Q_{mft} = a + b_1 PY_t + b_2 RP_{ft} + b_3 P_{ft} + b_4 S_{f(t-1)} + b_5 PR_{ft} + b_6 NFA_t \quad ...(5.30)$$

$$\text{Log } Q_{mft} = a + b_1 \log PY_t + b_2 \log RP_{ft} + b_3 \log P_{ft} + b_4 \log S_{f(t-1)} + b_5 \log PR_{ft} + b_6 \log NFA_t \quad ....(5.31)$$

$$Q_{mft} = a + b_1 PY + b_2 RP_{ft} + b_3 P_{ft} + b_4 S_{f(t-1)} + b_5 PR_{ft} + b_6 EA_t + b_7 NFA_t \quad ...(5.32)$$

$$\text{Log } Q_{mft} = a + b_1 \log PY_t + b_2 \log RP_{ft} + b_3 \log P_{ft} + b_4 \log S_{f(t-1)} + b_5 \log PR_{ft} + b_6 \log EA_t + b_7 \log NFA_t \quad ....(5.33)$$

$$Q_{mft} = a + b_1 RPf_t + b_2 PF_t + b_3 S_{f(t-1)} + b_4 EA_t + b_5 NFA_t + D + DS_{f(t-1)} \quad ....(5.34)$$

$$\text{Log } Q_{mft} = a + b_1 \log RP_{ft} + b_2 \log P_{ft} + b_3 \log S_{f(t-1)} + b_4 \log EA_t + b_5 \log NFA_t + D + DS_{f(t-1)} \quad ...(5.35)$$

$$Q_{mft} = a + b_1 PY_t + b_2 RP_{ft} + b_3 P_{ft} + b_4 S_{f(t-1)} + b_5 EA_t + b_6 NFA_t + D + DS_{f(t-1)} \quad ....(5.36)$$

$$\text{Log } Q_{mft} = a + b_1 \log PY_t + b_2 \log RP_{ft} + b_3 \log P_{ft} + b_4 \log S_{f(t-1)} + b_5 \log EA_t + b_6 \log NFA_t + D + DS_{f(t-1)} \quad ...(5.37)$$

$$Q_{mft} = a + b_1 Py_t + b_2 RP_{ft} + b_3 P_{ft} + b_4 S_{f(t-1)} + b_5 PR_{ft} + b_6 EA_t + D + DS_{f(t-1)} + DP \quad ....(5.38)$$

$$\text{Log } Q_{mft} = a + b_1 \log PY_t + b_2 \log RP_{ft} + b_3 \log P_{ft} + b_4 \log S_{f(t-1)} + b_5 \log PR_{ft} + b_6 \log EA_t + D + DS_{f(t-1)} + DP \quad .....(5.39)$$

$$Q_{mft} = a + b_1 PY_t + b_2 RP_{ft} + b_3 P_{ft} + b_4 S_{f(t-1)} + b_5 PR_{ft} + b_6 EA_t + b_7 NFA_t + D + DS_{f(t-1)} + DP \quad ....(5.40)$$

$$Q_{mft} = a + b_1 \log PY_t + b_2 \log RP_{ft} + b_3 \log P_{ft} + b_4 \log S_{f(t-1)} + b_5 \log PR_{ft} + b_6 \log EA_t + b_7 \log EA_t + b_8 \log NFA_t + D + DS_{f(t-1)} + DP \quad ...(5.41)$$

All the import demand equations from 5.2 to 5.41 as specified in Section 5.3.3. are estimated by the method of ordinary least squares and the best fitted form among these is selected for analysis on the basis of the maximum value of R^2 and the significance levels of the parameters.

5.4. Data Sources

Data relating to quantity of import of foodgrains and the import prices are obtained from U.N., 'Yearbook of International Trade Statistics', F.A.O, 'Trade Yearbook, and GOI., 'Bulletin on Food Statistics'. Per capita income is taken from GOI., 'Economic Survey'. Domestic wholesale prices of foodgrains are taken from R.B.I, 'Report on Currency and Finance'. The index of production of foodgrains in India is taken from 'Estimates of Area and Production of principal Crops in India', and 'Economic Survey'. Data relating to stocks of foodgrains and procurement of foodgrains are taken from G.O.I., 'Bulletin on Food Statistics'. Data relating to external assistance and net foreign assets are taken from G.O.I., 'Economic Survey'.

Data relating to the above set of variables expressed in index, are given in Table 5.1. The estimated exponential growth rates for these variables are given in Table 5.2 for the pre-Green Revolution period 1955-56 to 1966-67, for the post-Green Revolution period 1967-68 to 1986-87 and for the total period 1955-56 to 1986-87.

5.5. Analysis of Import Demand Functions for Imports of Foodgrains of India—Empirical Results

From the estimated values of the parameters of the import demand functions along with the values of 't', R^{-2} and F, it is found that the linear from gives a better fit to the data than the log linear form. Therefore, the linear forms of import demand functions are chosen for analysis. The regression results of the import demand functions for imports of foodgrains of India are presented in Table 5.3.

It is evident from Table 5.3 that in equation (1) per capita income (PY_t) is considered as an independent variable to explain the variation in the demand for imports of foodgrains of India. The variable explained 20.74 per cent of the variation in the demand for imports of foodgrains. The coefficient of the variable is found to be significant at one per cent level, but does not have the expected positive sign. In equation (2), the independent variable *viz.*, relative price (RP_{ft}) is considered. It is found to explain 27.12 per cent of the variation in the demand for imports of foodgrains. It has the expected negative sign and is found to be statistically significant at one per cent level.

Table 5.1 : Data used for Estimating the Parameters of the Import Demand Equations

	Q_{mft}	PY_t	RP_{ft}	P_{ft}	S_{ft}	PR_{ft}	EA_t	NFA_t
Period	Index of Quantity of Imports of Foodgrains (1970-71 = 100)	Percapita Income (1970-71 = 100)	Index of Relative Price (1970-71 = 100)	Index No. of Foodgrains Production in India (Triennium ending) (1969-70 = 100)	Stocks of Foodgrains (1970-71 = 100)	Procurement of Foodgrains (1970-71 = 100)	External Assistance (1970-71 = 100)	Net Foreign Assets (1970-71 = 100)
(1)	(2)	(3)	(4)	(5)	(6)	(7)	(8)	(9)
1954-55	..	..	..	..	1.67	..	..	..
1955-56	19.58	80.23	231.8	72.6	16.52	0.45	40.95	117.27
1956-57	39.00	82.93	152.0	75.9	5.74	3.39	65.63	88.51
1957-58	101.29	79.54	165.5	68.8	21.18	5.98	40.56	54.74
1958-59	88.79	84.42	127.9	82.1	16.34	20.43	61.95	49.24
1959-60	106.53	84.10	129.9	80.4	24.96	14.45	60.77	47.16
1960-61	141.48	88.29	133.1	86.1	50.27	6.09	127.31	39.45
1961-62	96.25	89.11	133.4	86.8	47.40	5.42	55.88	38.64
1962-63	100.25	88.46	133.2	83.8	40.93	8.47	84.55	38.35
1963-64	127.21	91.09	126.4	85.3	40.57	16.14	62.37	39.74
1964-65	184.52	96.05	105.5	94.3	18.31	45.49	93.92	32.45
1965-66	205.51	88.31	91.3	75.8	37.34	45.26	85.21	38.73
1966-67	285.27	87.15	99.9	71.1	39.86	50.34	197.73	62.17
1967-68	238.83	92.81	97.3	98.7	35.19	76.86	94.34	69.99

1968-69	156.82	93.09	114.2	97.3	71.63	72.01	120.27	74.55
1969-70	106.64	96.81	113.5	100.0	79.89	75.73	83.25	110.66
1970-71	100.00	100.00	100.00	112.9	100.00	100.00	100.00	100.00
1971-72	56.57	99.02	101.8	114.4	146.14	86.57	121.96	118.34
1972-73	22.03	95.35	79.0	102.3	61.75	95.03	88.75	117.99
1973-74	121.18	98.20	109.1	110.3	56.19	63.77	153.64	131.02
1974-75	154.23	97.50	84.9	104.3	49.01	107.90	219.30	133.58
1975-76	208.23	104.85	143.4	127.2	148.83	145.03	348.27	243.92
1976-77	179.43	193.05	172.6	115.7	340.39	111.40	168.84	420.82
1977-78	17.82	109.78	189.6	133.6	311.49	125.28	248.98	630.86
1978-79	6.47	113.31	234.4	139.3	308.80	156.21	306.56	754.18
1979-80	11.29	105.04	235.9	114.8	314.54	126.07	244.06	769.01
1980-81	11.84	110.35	368.3	137.5	210.77	146.50	504.91	719.99
1981-82	41.94	113.01	178.9	140.5	206.28	174.04	372.80	520.50
1982-83	36.96	114.10	148.1	135.8	229.26	176.86	387.03	618.73
1983-84	116.00	120.73	212.2	161.1	276.30	211.29	272.88	771.92
1984-85	26.27	122.47	161.3	154.3	403.59	227.09	615.83	936.86
1985-86	8.18	126.11	224.7	161.9	514.36	222.23	708.61	1,010.24
1986-87	3.55	129.42	211.8	154.0	509.52	177.73	761.74	1,047.29

Source:
1. G.O.I., *Bulletin on Food Statistics*, various issues.
2. G.O.I., *Economic Survey*, various issues.
3. G.O.I., CSO., *Statistical abstract*: India, various issues.
4. U.N., F.A.O., *Trade Year Book*, Rome, various issues.
5. U.N., *Yearbook of International Trade Statistics*, New York, various issues.
6. R.B.I., *Report on Currency and Finance*, various issues.
7. G.O.I., *Food Corporation of India*, various reports.
8. G.O.I, *Estimates of Area and Production of Principal Crops of India*, various issues.

Table 5.2 : Estimated Growth Rates[a] of the Variables considered for Import Demand Equations

Period	Quantity of Food-grains	Per capita income	Relative price	Production of Food-grains	Stocks of Foodgrains	Procurement of Food-grains	External Assistance	Net Foreign Assets.
1955-56 to 1966-67	16.78* (4.51)	1.48* (5.29)	-6.20* (5.07)	1.06 (1.62)	8.47*** (1.79)	30.24 (4.43)	6.30 (2.29)	-10.19 (4.98)
1967-68 to 1986-87	-13.92* (3.92)	1.60* (9.20)	4.95* (4.91)	2.51* (10.14)	12.13* (6.93)	6.13* (9.930	9.78* (7.86)	15.77* (13.76)
1955-56 to 1986-87	-6.01* (3.09)	1.33** (15.51)	1.53** (2.39)	2.55* (17.71)	11.10* (11.98)	14.09* (10.23)	8.30* (13.13)	11.60* (11.10)

Note: (a) Exponential growth rate.
Figures in brackets represent 't' values.
* Significant at one per cent level
** Significant at five per cent level.
*** Significant at ten per cent level.

Table 5.3 : Estimated Import Demand Functions for Imports of Foodgrains of India: 1955-56 to 1986-87
(Dependent Variable: Quantum of Imports of Foodgrains of India)

Equation Serial Number	Constant	Explanatory variables						
		PY_t	RP_{ft}	Pft	$S_{f(t-1)}$	PR_{ft}	EA_t	NFA_t
1.	2.	3.	4.	5.	6.	7.	8.	9.
1.	12.3572	-0.0141* (2.8074)	—	—	—	—	—	—
2.	6.9672	—	-0.0228* (3.3416)	—	—	—	—	—
3.	11.3692	-0.0083 (1.5583)	-0.0172** (2.2768)	—	—	—	—	—
4.	8.9650	—	0.0169** (2.2127)	-0.0279 (1.5463)	—	—	—	—
5.	6.0388	—	-0.0080 (0.9734)	—	-0.1087* (2.7037)	—	—	—
6.	14.7437	—	—	-0.1397** (2.2349)	—	0.3976 (1.5258)	—	—
7.	10.5310	-0.0054 (0.2236)	-0.0171** (2.1853)	-0.0104 (0.1285)	—	—	—	—
8.	4.1785	—	-0.0070 (0.8342)	0.0212 (0.7683)	-0.2474** (2.2717)	—	—	—
9.	8.7569	—	—	-0.05474 (0.9434)	-0.2851* (3.1192)	0.3443 (1.5028)	—	—
10.	4.4319	—	—	0.0041 (0.1401)	-0.3885* (3.6869)	—	0.0009*** (1.7083)	—

(Contd.)

Table 5.3 (Contd.)

Equation Serial Number	Explanatory variables			R^2	F-value	DW-statistic
	D	$D_{sf(t-1)}$	DP			
	10.	11.	12.	13.	14.	15.
1.	—	—	—	0.2074	7.8480	0.6030
2.	—	—	—	0.2712	11.1662	1.0598
3.	—	—	—	0.3276	7.0632	0.9947
4.	—	—	—	0.3268	7.0374	1.0365
5.	—	—	—	0.4180	10.4122	0.7583
6.	—	—	—	0.2716	5.4056	0.6804
7.	—	—	—	0.3280	4.5546	1.0086
8.	—	—	—	0.4292	7.0402	7.1998
9.	—	—	—	0.4594	7.9317	0.7653
10.	—	—	—	0.4709	8.3082	0.6057

(Contd.)

Table 5.3 (Contd.)

Equation Serial Number	Constant	Explanatory variables						
		PY_t	RP_{ft}	P_{ft}	$S_{f(t-1)}$	PR_{ft}	EA_t	NFA_t
1.	2.	3.	4.	5.	6.	7.	8.	9.
11.	5.6345	—	-0.0075	-0.0002	-0.3388*	—	0.0009***	—
			(0.9249)	(0.0073)	(2.8599)		(1.7357)	
12.	7.8861	—	-0.0143	-0.0131	-0.2634***	—	—	—
			(1.5057)	(0.3733)	(2.4437)			
13.	7.6827	—	-0.0176***	0.0063	-0.6703**	—	—	—
			(1.8265)	(0.1695)	(2.1214)			
14.	8.7016	—	-0.0163	-0.0162	-0.5747***	—	0.0007	—
			(1.6966)	(0.3927)	(1.7797)		(1.2126)	
15.	-20.1078	0.0750**	0.0155	-0.2376**	-0.3336***	0.4267	—	-0.0010
		(2.3582)	(1.2919)	(2.4849)	(1.8665)	(1.6372)		(1.3972)
16.	-16.6049	0.0652***	0.0137	-0.2123***	-0.3244***	0.3967	0.0003	-0.0010
		(1.7455)	(1.0841)	(1.9544)	(1.7801)	(1.4656)	(0.5189)	(1.4166)
17.	6.1885	—	-0.0134	0.0173	-0.6042***	—	0.0009	-0.0015***
			(1.4485)	(0.4035)	(1.9667)		(1.6672)	(1.9216)
18.	-13.0268	0.0498	-0.0992	-0.992	-0.6604**	—	0.0005	-0.0016**
		(1.4019)	(0.2957)	1.0643)	(2.1732)		(0.8825)	(2.0852)
19.	14.0220	0.0540	-0.0006	-0.1348	-1.1509*	0.3999***	0.0007	—
		(1.6899)	(0.0572)	(3.6718)	(1.7118)	(1.1575)		
20.	-13.9439	0.0574***	0.0045	-0.1639***	-1.0715*	0.5575**	0.0005	-0.0014
		(1.8204)	(0.4056)	(1.7593)	(3.4143)	(2.1584)	(0.7808)	(1.3223)

(Contd.)

Table 5.3 (Contd.)

Equation Serial Number	Explanatory variables			R^2	F-value	DW-statistic
	D	$D_{sf(t-1)}$	DP			
	10.	11.	12.	13.	14.	15.
11.	—	—	—	0.4872	6.4128	0.7578
12.	2.8617 (1.5393)	—	—	0.4760	6.1307	0.9405
13.	0.7164 (0.2972)	0.4383 (1.3672)	—	0.5111	5.4364	1.0286
14.	1.4524 (0.5893)	0.2613 (0.7474)	—	0.5383	4.8573	1.0294
15.	—	—	—	0.5689	5.4975	0.8646
16.	—	—	—	0.5736	4.6129	0.8619
17.	1.8931 (0.8045)	0.5629 (1.5319)	—	0.5998	5.1394	1.1832
18.	2.0933 (0.9056)	0.4922 (1.3522)	—	0.6313	4.9235	1.1390
19.	6.7528** (2.3849)	0.6447*** (1.95.41)	-0.8253* (3.3711)	0.7150	6.1338	1.4884
20	4.3334 (1.3121)	0.8506** (2.3680)	-0.5557*** (1.7675)	0.7373	5.8924	1.4575

Note: Figures in brackets represent 't' values.
* Significant at one per cent level
** Significant at five per cent level.
*** Significant at ten per cent level.

In equation (3), the two independent variables *viz.*, per capita income (PY_t) and relative price (RP_{ft}) are considered together to explain the variation in the demand for imports of foodgrains. In this equation, the coefficient of per capita income (PY_t), is found to be not significant, while the coefficient of relative price (RP_{ft}) is found to be statistically significant at 5 per cent level. The two variables together explained 32.76 per cent of the variation in the demand for imports of foodgrains. Thus, the inclusion of relative price (RP_{ft}) variable in the equation has increased the explanatory power of equation from 20.74 per cent [equation (1)] to 32.76 per cent [equation (3)].

In equation (4), the two independent variables relative price (RP_{ft}) and production of foodgrains (P_{ft}) are considered together to explain the variation in the demand for imports of foodgrains. Of these two variables, the coefficient of relative price (RP_{ft}) alone is found to be significant at 5 per cent level, but does not have the expected negative sign. Together, these two variables explained 32.68 per cent of variation in the demand for imports of foodgrains.

In equation (5), the two independent variables *viz.*, relative price (RP_{ft}) and stocks of foodgrains lagged by one year [$S_{f(t-1)}$] are considered together to explain the variation in the demand for foodgrains. The coefficients of both the variables have the expected signs. Among these, the coefficient of $S_{f(t-1)}$ alone is found to be significant at one per cent level. Together, they explained 41.80 per cent of the variation in the demand for imports of foodgrains, while RP_{ft} combined with PY_t explained only 32.76 per cent of the variation in the import demand for foodgrains [See equation (3)].

In equation (6), the two independent variables *viz.*, production of foodgrains (P_{ft}) and procurement of foodgrains (PR_{ft}) are considered together to explain the variation in the demand for foodgrains. Of these two, the coefficient of production of foodgrains (P_{ft}) alone is found to be significant at 5 per cent level and has the expected negative sign. Together, they explained 27.16 per cent of the variation in the demand for imports of foodgrains.

In equation (7), we have taken the three independent variables *viz.*, per capita income (PY_t), relative price (RP_{ft}) and production of foodgrains (P_{ft}) as explanatory variables to explain the variation in the demand for imports of foodgrains. Among the three variables, the coefficient of relative price (RP_{ft}) alone is found to be significant at 5 per cent level and has the expected negative sign. The variables included in this equation together explained 32.80 per cent of variation in the demand for imports of foodgrains.

In equation, (8), the independent variables *viz.*, relative price (RP_{ft}), production of foodgrains (P_{ft}) and stocks of foodgrains ($S_{f(t-1)}$) are considered together to examine the influence of these three variables on the demand for imports of foodgrains. Among these, the coefficient of the variables $S_{f(t-1)}$ alone is found to be significant at five per cent level with the expected negative sign. The three variables together explained 42.99 per cent of the variation in the imports of foodgrains.

In equation (9), we have taken production of foodgrains (P_{ft}), stocks of foodgrains [$S_{f(t-1)}$] and procurement of foodgrains (PR_{ft}) as explanatory variables to explain the variation in the demand for imports of foodgrains. Among these, the coefficient of the variable $S_{f(t-1)}$ alone is found to be significant at one per cent level with the expected negative sign. The three variables together explained 45.94 per cent of the variation of the demand for imports of foodgrains.

In equation (10), the three independent variables *viz.*, production of foodgrains (P_{ft}), stocks of foodgrains [$S_{f(t-1)}$] and external assistance (EA_t) are considered together to explain the variation in the demand for foodgrains. Among these, the coefficients of the variable $S_{f(t-1)}$ is found to be statistically significant at one per cent level, with the expected negative sign and the coefficient of the variable EA_t is found to be statistically significant at ten per cent level, with the expected positive sign. The coefficient of P_{ft} does not have the expected negative sign, but is not found to be statistically significant. These three variables together explained 47.09 per cent of the variation in the import demand for foodgrains. Thus, there is an increase in the explanatory power of equation (10) to 47.09 per cent when the variable EA_t is considered along with P_{ft} and $S_{f(t-1)}$.

In equation (11), the four independent variables *viz.*, relative price (RP_{ft}), production of foodgrains (P_{ft}), stocks of foodgrains [$S_{f(t-1)}$] and external assistance (EA_t) are considered together as explanatory variables to examine the variation in the demand for imports of foodgrains. Among these, the coefficient of the variable [$S_{f(t-1)}$] is found to be statistically significant at one per cent level, with the expected negative sign and the coefficient of the variable EA_t is found to be statistically significant at ten per cent level, with the expected positive sign. The other two variables *viz.*, relative price (RP_{ft}) and production of foodgrains (P_{ft}) are not found to be statistically significant. The four variables considered in this equation increased the explanatory power of the equation to 48.72 per cent.

In equation (12), the dummy variable 'D' is considered along with the other independent variables *viz.*, relative price (RP_{ft}), production of foodgrains (P_{ft}) and stocks of foodgrains [$S_{f(t-1)}$] in explaining the variation in the demand for imports of foodgrains. Among these variables, the coefficient of the variable $S_{f(t-1)}$ alone is found to be statistically significant witn the expected negative sign. The coefficients of the other variables are not found to be statistically significant. The explanatory power of equation (12) is found to be 47.60 per cent.

In equation (13), we have taken the independent variables RP_{ft}, P_{ft}, $S_{f(t-1)}$ along with the 'intercept' dummy variable 'D' and the slope dummy variables $DS_{f(t-1)}$ to explain the variation in the demand for imports of foodgrains. Among these the coefficients of the variables RP_{ft} and $S_{f(t-1)}$ alone are found to be statistically significant with expected negative signs. There is an increase in the explanatory power of equation (13) to 51.11 per cent.

In equation (14), we have considered the independent variables, *viz.*, relative price (RP_{ft}), production of foodgrains (P_{ft}), stocks of foodgrains [$S_{f(t-1)}$], external assistance (EA_t) along with the intercept dummy variable 'D' and slope dummy variables '$DS_{f(t-1)}$', to explain the variation in the import demand for foodgrains. Among these variables, the coefficient of the variables stocks of foodgrains ($S_{f(t-1)}$ alone is found to be statistically significant at ten per cent probability level, with the expected negative sign. The coefficients of the other variables are not found to be statistically significant. There is an increase in the explanatory power of equation (14) to 53.83 per cent.

In equation (15), we have considered per capita income (PY_t), relative price (RP_{ft}), production of foodgrains (P_{ft}), stocks of foodgrains [$S_{f(t-1)}$], procurement of foodgrains (PR_{ft}) and net foreign assets (NFA_t) as the independent variables, to explain the variation in the demand for imports of foodgrains of India. Among these, coefficients of the variables PY_t, P_{ft} and $S_{f(t-1)}$ alone are found to be statistically significant, with expected signs. The variables considered to this equation together explained 56.89 per cent of the variation in the demand for imports of foodgrains.

In equation (16), we have taken the independent variables *viz.*, per capita income (PY_t), relative price (RP_{ft}), production of foodgrains (P_{ft}), stocks of Foodgrains [$S_{f)t-1)}$], procurement of foodgrains (PR_{ft}), external assistance (EA_t) and net foreign assets (NFA_t) to explain the variation in the demand for imports of foodgrains. Among these, the coefficients of the variables PY_t, P_{ft} and $S_{f(t-1)}$ alone are found to be statistically significant at ten per cent level, with expected signs. There is an increase in the explanatory power of Equation (16) to 57.36 per cent.

In the equation (17), we have considered the independent variables *viz.*, relative price (RP_{ft}), production of foodgrains (P_{ft}), stocks of foodgrains ($S_{f(t-1)}$), external assistance (EA_t), net foreign assets (NFA_t) along with the 'intercept' dummy variable 'D' and 'slope' dummy '$DS_{f(t-1)}$' in explaining the variation in the demand for imports of foodgrains. Among these, the coefficient of the variable $S_{f(t-1)}$ is found to be statistically significant at ten level, with the expected negative sign. The coefficient of the variable NFA_t is found to be statistically significant at ten per cent level, but does not have the expected positive sign. The variables considered in this equation together explained 59.98 per cent of the variation in the demand for imports of foodgrains of India.

In equation (18), we have taken the independent variables *viz.*, PY_t, RP_{ft}, P_{ft}, $S_{f(t-1)}$, EA_t and NFA_t, along with the dummy variables to explain the variation in the import demand for foodgrains in India. Among these variables, the coefficient of the variables $S_{f(t-1)}$ is found to be significant at five per cent level, with the expected negative sign. The coefficient of the variables NFA_t is found to be significant at five per cent level, but does not have the expected positive sign. The variables considered in this equation together explained 63.13 per cent of the variation in the demand for imports of foodgrains.

In equation (19), we have considered the independent variables *viz.*, PY_t, RP_{ft}, P_{ft}, $S_{f(t-1)}$, PR_{ft} and EA_t along with the dummy variables. The coefficient of the variables $S_{f(t-1)}$ has the expected negative sign and is found to be statistically significant at one per cent probability level. The coefficient of the variable PR_{ft} is found to be significant at ten per cent level, but does not have the expected negative sign. The coefficients of the dummy variables *viz.*, D, $DS_{f(t-1)}$ and DP are found to be statistically significant. The coefficients of the other variables, *viz.*, PY_t, RP_{ft}, P_{ft} and EA_t, however, are not found to be statistically significant. The variables considered in this equation, together explained 71.50 per cent of the variation in the demand for imports of foodgrains of India.

In equation (20), all the independent variables, *viz.*, PY_t, RP_{ft}, P_{ft}, $S_{f(t-1)}$, P_{ft}, EA_t, NFA_t and the dummy variables D, $DS_{f(t-1)}$ and DP are considered together in order to examine the influence of these variables on the demand for imports of foodgrains of India. Among these, the coefficients of the variables PY_t, P_{ft}, $S_{f(t-1)}$ and PR_{ft} are found to be significant. However, the coefficient of the variable PR_{ft} does not have the expected negative sign. The coefficients of RP_{ft} and NFA_t do not have the expected signs, but are not found to be statistically significant. The coefficient of the variable EA_t has the expected positive sign, but is not found to be statistically significant. The variables considered in this equation together explained 73.73 per cent of the variation in the demand for imports of foodgrains of India.

Out of the twenty equations estimated and presented in Table 5.3, equation (20) is considered to be the best fit as it has the highest explanatory power. In this equation, the independent variables *viz.*, PY_t, RP_{ft}, P_{ft}, $S_{f(t-1)}$, PR_{ft}, EA_t, NfA_t, along with the dummy variables together explained 73.73 per cent of the variation in the demand for imports of foodgrains of India. From equation (20), it may be inferred that a one unit increase in per capita income leads to 0.06 units increase in the imports of foodgrains. The coefficient of production of foodgrains has the expected negative sign and is found to be statistically significant at 10 per cent level. It indicates that a one unit increase in the production of foodgrains leads to 0.16 units decline in the imports of foodgrains.[24]

The coefficient of the variable $S_{f(t-1)}$ is found to be significant at one per cent probability level with expected negative sign. From

24. The elasticity of foodgrains imports with respect of domestic production is found to be greater than unity (-4.8970), indicating that, on an average, a one per cent increase in foodgrain production leads to 4.90 per cent decline in foodgrain imports.

this it may be inferred that a one unit increase in the stocks of foodgrains in year (t-1) leads to 1.07 units decline in the imports of foodgrains in year 't', during the first period 1955-56 to 1974-75. During the second period, 1975-76 to 1986-87, the coefficient of the variables $S_{f(t-1)}$ is found to be -0.2209.[25] From this it may be inferred that a one unit increase in the stocks of foodgrains in year (t-1) leads to a 0.2209 units decline in the imports of foodgrains in year 't' during the second period, 1975-76 to 1986-87.

The coefficient of the variable PR_{ft} is found to be significant at 5 per cent level in equation (20). The positive sign of the coefficient may be due to the need to import foodgrains in order to maintain the required level of procurement of foodgrains. From this, it may be inferred that a one unit increase in the procurement of foodgrains leads to 0.5575 units increase in the imports of foodgrains during the first period 1955-56 to 1974-75. During the second period 1975-76 to 1986-87, the coefficient of the variable is found to be 0.0018.[26] From this it may inferred that a one unit increase in the procurement of foodgrains leads to 0.0018 units increase in the imports of foodgrains during the second period.

5.6 Comparison of the Results with Earlier Studies

In the estimated import demand function for the imports of Food and Live Animals of India under SITC Section 'O', Jayarama[27] found

25. This is arrived at by adding the coefficients of the variable $S_{f(t-1)}$ and of the dummy variable $DS_{f(t-1)}$. This gives the coefficient for $S_{f(t-1)}$ for the second period *viz.*, 1975-76 to 1986-87. It may be noted here that there is a decline in the magnitude of the coefficient from the first period to the second period. Dummy variable 'D' takes the value one from the year 1975-76 and zero for the earlier years. Hence, the period 1955-56 to 1974-75 is treated as the first period and the period 1975-76 to 1986-87 is treated as the second period.
26. This is arrived at by adding the coefficients of the variables PR_{ft} and the dummy variable 'DP'. This gives the coefficient for PR_{ft} for the second period, *i.e.*, 1975-76 to 1986-87. It may be noted here that there is a substantial decline in the magnitude of the coefficient from the first period to the second period.
27. T.K. Jayaraman, "Import Demand Function of India", *op. cit.*, pp. 362-372, Jayaraman arrived at the following results:

Intercept	$\frac{PM}{PW}$	$\frac{GNP}{PW}$	$\frac{NFA}{PW}$	D_1	T	R^2	F	DW Statistic
5.0006	-3.605	9.029	-0.133	2.749	-0.619	0.9400	15.66	2.5179
(23.109)	(-3.847)	(3.697)	(-1.781)	(8.126)	(-5.814)			

Where, PM = price of imports, PW = domestic wholesale price indeed, GNP = gross national product, NFA = net foreign assets in current prices, D_1 = dummy variable and T = time period.

the coefficients of the explanatory variables *viz.*, relative price, gross national product, net foreign assets and dummy variables (for devaluation) to be statistically significant. Imports of Food and Live Animals of India are found to have high price and income elasticities. The variable net foreign assets does not have the expected positive sign, though its coefficient is found to be significant. However, the study has not considered imports of foodgrains separately.

In the estimated function for Food imports of India, P.C. Verma[28] found that the coefficient of per capita food production variable is highly significant. It also has the expected sign. A one per cent increase in per capita food production reduces food imports by 3.3 per cent. Food stocks variable is also found to be significant, but does not have expected sign. However, the study of the author relates to the period 1960-61 to 1969-70 only.

In the estimated import demand function for India's foodgrains imports, Basudeb Biswas and Rati Ram[29] have found that the

---

28. P.C. Verma, "Import Structure of India, Bangladesh and Pakistan", *op. cit.*, pp.498-570. The estimated import demand equation is given as follows:

$$\text{Log } M_{fd} = 9.988 - 3.317 \text{ Log } FPd + 0.0006FS$$
$$(-4.701) \qquad (2.444)$$
$$R^2 = 0.78 \qquad DF = 7$$

where,

M_{fd} = food imports at constant prices

FPD = per capita food production index and

FS = food stock changes over previous year

29. B. Biswas and R. Ram, *op. cit.*, p. 16. They arrived at the following results:

Y_t	P_{mt}	P_{dt}	PRD_t	GS_t	FER_t	D
1.878	0.075	0.296	-2.698	-0.429	-1.064	0.785
(0.54)	(0.09)	(0.27)	(1.36)	(2.43)	(3.87)	(1.54)

R^2	DW-statistic	Standard error of the Regression
0.815	1.93	0.415

where,

Y_t = GNP for year 't' at 1961 prices (in million rupees),

P_{mt} = Price of imported foodgrains measured by the C & F cost (in rupees) per metric ton, in year 't',

P_{dt} = Index number of (domestic) wholesale prices of foodgrains (1961-62 = 100) for year 't',

PRD_t = Domestic production of foodgrains in year 't' (million metric tons),

GS_t = The opening level of the Central stocks of foodgrains for the year t (million metric tons) and

FER_t = Foreign exchange reserves in the beginning of year t.

coefficient of the Government stock variable has the expected negative sign and is found to be significant at one per cent level. The signs of the coefficients of income (Y) and domestic foodgrains prices (P_d) are positive as expected, but their t-statistics are quite low. The coefficient of the import price variable has the wrong sign, but is not found to be statistically significant. The dummy for the buffer stock years has the expected sign, but moderate though not significant t-statistic.

The authors found that the central foodgrain stocks clearly had a strong negative effect. The effect of the income and domestic foodgrain price level is found to be relatively weak. The effect of import price is also found to be weak. The coefficient of foreign exchange reserves suggests that the reserve level is not an important consideration, and historically low foreign exchange reserves have been associated with high imports. Thus, the authors observed strong effect of central stocks, modest effect of domestic production and small effects of imcome and world price. The analysis of the authors covers the 25-year period 1951-75.

S.K. Patel[30] has estimated import demand function for food

30. S.K. Patel, *op. cit.*, p.79 and p.89. Results of the estimated Food import demand equations are as follows:

Regression results - Imports of Food(MF) - Pre-devaluation period

Constant	Food Prices	Population	Agricultural Production Index	Seasonal Dummies D_1	D_2	D_3
3422.7	14507.0 (5.56)	-348.8 (1.44)	4.09 (0.20)	1477.0 (0.59)	1778.2 (0.58)	186.7 (0.06)

R^2	F-value	d	d4
0.88	21.84	2.76	2.21

Regression Results - Imports of Food(MF) - Post-devaluation period

Constant	Food Prices	Population	Agricultural Production Index	Seasonal Dummies D_1	D_2	D_3
7.448.0	2459.6 (0.97)	-1465.0 (5.71)	13.44 (0.55)	3107.2 (0.71)	2153.7 (0.41)	3055.3 (0.61)

R^2	F-value	d	d4
0.82	16.46	1.60	1.60

imports of India based on Standard Internatinal Trade Classification (SITC). In the pre-devaluation period the author found that food price variable alone is significant. In the post-devaluation period population variable is found to be significant but with a perverse sign. The analysis of the period relates to 1958-59 to 1976-77 and to the SITC (one-digit) classification. The author has not analysed foodgrain imports separately.

In the estimated wheat import demand function, Gulati[31] found that except the price variable, the coefficients of all other variables *viz.*, closing stocks of foodgrains in the previous period, wheat output in the current period and external aid, are found to be significant at 5 per cent level. The estimated equation explains 91 per cent of the variation in wheat imports. However, he has not considered foodgrain imports as a whole, but only wheat imports.

Ramesh Babu[32] found that India's Food imports are determined to a greater extent by the availability of external assistance, including the tied-aid under PL 480 agreement. Domestic economic activity

---

31. A. Gulati, Agricultural Price Policy in India, *op-cit.*, pp.110-111. Results of the estimated wheat import demand equation are as follows:

MW(t) = 6372.28-0.179 CSAF (t-1)-0.179 QW (t) -14.185 LPMW (t)
(3.57) (-2.56) (-2.77) (0.98)
-0.332 -1.1081 -0.444

+ 3.624 EAU (t)
(4.20)
1.12 · $R^2 = 0.91$, DW = 1.22

where,

MW(t) = Imports of wheat in year 't' in thousand tonnes
CSAF(t-1) = Closing stock of all foodgrains with the Government as on 31st Dec. of the previous calendar year in thousand tonnes.
LPMW(t) = Ratio between landed price of imported wheat and unit value index of all imports, 1970-71 = 100 and'
EAU(t) = External Aid utilised during the financial year t, Rs. 10 million.

32. S. Ramesh Babu, *op. cit.*, p.134. Results of the estimated Food and Live Animals import demand equation are as follows:

Constant	RP	GDP	EA	FPI	R-2	DW
8.42	-0.06	-1.02	1.39*	-	0.65	1.53
	(0.22)	(1.49)	(5.89)			

where,

RP_t = Relative Price (Price of imports to domestic prices)
GDP_t = Gross domestic product of India
EA_t = External assistance including PL 480 and
FPI = Index of Food Production.

including production of food and relative prices do not seem to be having any influence on the import of Food and Live Animals of India. Explanatory power of the equation is found to be 65 per cent. However, his study relates to imports of Food and Live Animals under RITC. He has not considered foodgrain imports separately.

The Present study analyses the import demand for foodgrains separately. It considers per capita income, relative price, foodgrain production, stocks of foodgrains lagged by one year, procurement of foodgrains, external assistance, net foreign assets and dummy variables as explanatory variables for the demand for imports of foodgrains of India. Among these, the coefficients of the variables, per capita income, foodgrains production, stocks of foodgrains lagged by one year and procurement of foodgrains are found to be statistically significant. The analysis has been carried out for the period 1955-56 to 1986-87. The study indicates that imports of foodgrains of India are determined by per capita income, domestic production of foodgrains, stocks of foodgrains in the previous period and procurement of foodgrains. Imports of foodgrains are also influenced by buffer stock policy and procurement policy of the Government. The impact of Government's policy on imports of foodgrains is found to be less in the second period. 1975-76 to 1986-87 compared to the first period 1955-56 to 1974-75.

5.7. Summary

The major factors determining imports of foodgrains of India considered in this Chapter are: per capita income (PY_t), relative price (RP_{ft}), foodgrains production (P_{ft}), stocks of foodgrains lagged by one year $S_{f(t-1)}$, procurement of foodgrains (PR_{ft}), external assistance (EA_t) and net foreign assets (NFA_t). In addition, dummy variables have been used to capture the effects of policy changes in respect of buffer stocks and procurement. The significance of these variables and the nature and extent of the influence of these variables on the imports of foodgrains are analysed in this Chapter.

Among the variables considered, the coefficients of per capita income, production of foodgrains, sticks of foodgrains lagged by one year and procurement of foodgrains are found to be statistically significant. A one unit increase in the per capita income leads to 0.06 units increase in the demand for imports of foodgrains. On the other hand, a one unit increase in the production of foodgrains leads to 0.16 units decline in the imports of foodgrains. Similarly, increase in the stocks of foodgrains in the previous year leads to decline in the imports of foodgrains in the current year. On the other hand, increase in the procurement of foodgrains is found to lead to increase in the imports of foodgrains, when procurement of foodgrains is not found to be at required level.

From this analysis, it may be concluded that per capita income, domestic production of foodgrains, stocks of foodgrains, procurement of foodgrains and Government policies are found to be the major factors determining the demand for imports of foodgrains of India during the period 1955-56 to 1986-87.

CHAPTER 6

SUMMARY AND CONCLUSIONS

6.1. Introduction

COUNTRIES with a relatively large agricultural sector depend more on agricultural growth to sustain their development than do the others. Indian economy is basically agricultural in nature. Agriculture and allied activities contributed about 33 per cent to the Gross Domestic Product (GDP) of India in 1987-88. Agricultural sector continues to provide livelihood to about 75 per cent of the population of the country. Agricultural and allied exports accounted for about 19 per cent of the total exports of India in 1987-88. An increase in agricultural output is essential for providing the necessary food for the growing population, raw materials for the industry and markets for industrial products. Thus, agriculture is an important segment of the Indian economy.

About one-quarter of economic activity in India is involved in the production of foodgrains. Foodgrain crop plays a crucial role in the agricultural economy of India. Foodgrains produced in India are classified into three major groups, *viz.*, (*a*) cereals, (*b*) coarse grains/millets and (*c*) pulses. India's main cereal crops are: rice and wheat. Jowar, bajra barley and maize constitute the important coarse grains of the country. Pulses consist of: redgram, blackgram, bengalgram, tur, arhar, masur, mung and other minor crops.

The dietary surveys conducted by the Indian Council of Medical Research (ICMR) indicate the importance of Cereals in the Indian diet for the different social classes. Cereals account for about two-thirds of India's total calorie intake and somewhat smaller proportion of the protein content of the average diet. Of the cereals, roughly one-half is rice, one-sixth is wheat and the remaining one-third consists of millets. The importance of pulses in the Indian economy is on account of two factors; (*a*) pulses offer an effective solution to the 'protein problem', and (*b*) they have a useful role to play in replenishing soil fertility as they add more nitrogen to the soil than fertilisers.

Food supply plays an important role in economic development. This fact is emphasised in most of the models of economic growth. The classical economists have assigned a central place to food in their growth models, which are relatively more relevant to the growth problems of the developing countries. Food surplus is, thus, of strategic importance to the economic development of the less developed countries. From this, it follows that highest proirity should be given to the increase in food production in the plans for economic development of the less developed countries. Food sector plays a crucial role in the developed of the Indian economy also.

The food problem in India is essentially a foodgrains problem, primarily because of the importance of foodgrains in the Indian diet. As a nation, India is short of food both in absolute terms as well as in comparative terms. Shortfall in calories is treated as under-nutrition. Mal-nutrition means a nutritionally unsound composition of the foods from which the calories are derived. about 250 million people of India are either under-nourished or mal-nourished or both. This gives, in capsule form, the nature of the food problem in India, apart from the intensified problems arising out of shortfalls in production. The long term food problem involves questions of production, availability of domestic supplies, distribution over time between States, effective demand and its relation to income and employment and of the needs of the vulnerable groups of the society.

International trade is an activity of strategic importance in the development process of a developing economy like India. The import content of a development programme would vary from country to country depending on the pattern of resource allocation and also on the choice of techniques of production. Economic growth has an effect on imports and imports in their turn influence economic growth, which involves a process of raising the productive capacity of the economy.

Persistent balance of payments difficulties have constrained and sometimes even frustrated India's effolts to generate self-sustained economic growth. 'Efficient import-management' has become necessary to bring about adjustment in the balance of payments not only in the short run but also in the long run. Consistent with the availability from within the country, it would, therefore, be necessary to evolve a pragmatic import strategy which would meet the essential import requirements of agriculture, industry, infrastructure and other sectors of the economy. It is necessary that a supportive import policy is pursued over a period of time.

Before the beginning of the First World War, India enjoyed a small export surplus in respect of foodgrains. In the year 1921, India became a net importer of foodgrains. Imports of foodgrains came to be built into the whole system of food management in India and became, henceforth, a permanent feature of India's food economy. The net imports of Cereals of India increased from 1.6 million tonnes in 1944 to a high level of 4.7 million tonnes in 1951, but declined to 3.9 million tonnes in 1952 and 2.0 million tonnes in 1953.

Imports of Food and Live Animals Chiefly for Food, accounted for 4.34 per cent of the total imports of India in 1985-86. Cereals and Cereal preparations constitute an important part of imports of Food and Live Animals Chiefly for Food, and accounted for 12.9 per cent of the imports of Food and Live Animals. Chiefly for Food in 1985-86. Foreign trade in foodgrains is a Government of India monopoly. The Government of India, acting through the Food Corporation India (FCI), decides on how much to import what types of grain.

Imports of foodgrains have become necessary to meet the deficiency in the requirement of foodgrains in the Indian economy. However, in spite of the important role played by foodgrain imports in the Indian economy, there has been a relative neglect of the study of foodgrain imports in relation to the food problem of India. Hence, an attempt is made in the present study to analyse the import of foodgrains of India in relation to the food problem of India.

The specific objectives of the study are as follows:

(*a*) to analyse the growth of the foodgrain economy of India in terms of area, production and yield per hectare of foodgrains for the country as a whole;

(*b*) to study the trends in the prices, per capita net availability, procurement, public distribution and stocks of foodgrains as well as the targets and shortfalls in foodgrain production in the various Five-Year Plans of India;

(*c*) to examine the trends in the imports of foodgrains of India, the pattern and direction of such imports and the import availability ration for foodgrains; and

(*d*) to identify the factors which determine the demand for imports of foodgrains of India and to suggest policy measures.

6.2. Major Findings of the Study

The major conclusions which have emerged from the present study are as follows:

6.2.1. *Foodgrains Economy of India :*

(*a*) Area under foodgrains in India increased from 110.56 million hectares in 1955-56 to 115.30 million hectares in 1966-67, *i.e.*, at a rate of 0.52 per cent per annum, during the pre-Green Revolution period. During the same period, production of foodgrains in India increased from 69.3 million tonnes to 74.2 million tonnes, *i.e.*, at a rate of 1.06 per cent per annum, though not significant. Similarly, yield per hectare of foodgrains in India increased from 605 kg in 1955-56 to 644 kg in 1966-67, *i.e.*, at a rate of 0.93 per cent per annum, though not significant. Thus, during the pre-Green Revolution period, increase in the production of foodgrains in not fund to be statistically significant.

(*b*) Area under foodgrains in India increased from 121.42 million hectares in 1967-68 to 127.20 million hectares in 1986-87, *i.e.*, at a rate of 0.31 per cent per annum, during the post-Green Revolution period. During the same period, production of foodgrains in India increased from 95.1 million tonnes to 143.4 million tonnes, *i.e.*, at a rate of 2.51 per cent per annum. Similarly, yield per hectare of foodgrains in India increased from 783 kg in 1967-68 to 1,128 kg in 1986-87, *i.e.*, at a rate of 2.20 per cent per annum. Thus, during the post-Green Revolution period, increase in the production of foodgrains is found to be significant and is found to be mostly due to increase in the yield per hectare of foodgrains.

(*c*) On the whole, during the period 1955-56 to 1986-87, area under foodgrains in India increased from 110.56 million hectares to 127.20 million hectares *i.e.*, at a rate of 0.49 per cent per annum. During the same period, production of foodgrains in India increased form 69.3 million tonnes to 143 million tonnes, *i.e.*, at a rate of 2.55 per cent per annum, while yield per hectare of foodgrains in India increased from 605 kg to 1,128 kg., *i.e.*, at a rate of 2.13 per cent per annum. Thus, during the period 1955-56 to 1986-87, increase in the production of foodgrains is found to be mostly associated with increase in the yield per hectare of foodgrains.

(*d*) Wide fluctuations in agricultural production have become a striking element of the Indian economy. Production of foodgrains in India showed considerable fluctuations, which emnate from fluctuations in area and yield. The coefficient of variation is found to be 4.95 per cent for area under foodgrains, 25.12 per cent for production of foodgrains and 2.11 per cent for yield per hectare of foodgrains during the period 1955-56 to 1986-87. Thus, the instability in the production of foodgrains in India seems to be mostly arising from instability in the yield per hectare of foodgrains. The coefficient of variation is found to be 2.55 per cent for area, 8.37 per cent for production and 7.26 per cent for yield per hectare of foodgrains during the pre-Green Revolution period 1955-56 to 1966-67. The coefficient of variation is found to be 2.57 per cent for area, 16.23 per cent for production and 14.18 per cent for yield per hectare of foodgrains during the post-Green Revolution period 1967-68 to 1986-87. Thus, the instability in foodgrains production during the pre-Green Revolution period as well as during the post-Green Revolution period is found to be associated with the instability in the yield per hectare of foodgrains. Further, the coefficient of variation is found to be increasing in respect of both production and yield per hectare of foodgrains during the period 1955-56 to 1986-87, indicating the increase in the degree of instability. In addition, the coefficient of variation is found to be higher in the post-Green Revolution period compared to be pre-Green Revolution period, in the case of both production of foodgrains as well as yield per hectare of foodgrains, indicating the greater instability in production and yield of foodgrains in the post-Green Revolution period.

(*e*) During the period 1955-56 to 1986-97, changes have taken place, to some extent, in the cropping pattern of foodgrains in India, with rice, wheat and maize showing increase and jowar, small millets and barley showing decline. Area under rice accounted for 28.5 per cent of the total area under foodgrains. The relative share of area under wheat increased substantially from 11.2 per cent in 1955-56 to 17.9 per cent in 1986-87. These two crops together accounted for 39.7 per cent in 1955-56, and 50.7 per cent in 1986-87, of the total area under foodgrains. Area under total cereals accounted for 77.7 per cent of the total area under foodgrains in 1955-56 and increased to 81.8 per cent in 1986-87, while the share of area under total pulses in total foodgrains showed a decline from 22.3 per cent in 1955-56 to 18.2 per cent in 1986-87, indicating changes in the cropping pattern of foodgrains in India.

(*f*) There are variations in the growth rates of production of different foodgrain crops. During the period 1955-56 to 1986-87, rice, wheat, jowar, bajra, maize and ragi showed positive and significant growth rates, while small millets and barley showed negative and significant growth rates. Wheat production showed the highest growth rate of 6.11 per cent per annum followed by rice with 2.49 per cent, maize with 2.37 per cent, ragi with 1.63 per cent, bajra with 1.36 per cent and jowar with 1.04 per cent per annum. Among these, the growth rate of wheat production (6.11 per cent per annum) alone is found to be higher than the growth rate of 2.55 per cent per annum of total foodgrains production in India.

Rice is the major foodgrain produced in India, accounting for 42.26 per cent, followed by wheat with 30.89 per cent of the total foodgrains produced in India in 1986-87. Together, rice and wheat accounted for 73.15 per cent of the total foodgrains production in 1986-87. There are some changes in the pattern of production of foodgrains in India during the period 1955-56 to 1986-87. There is a rise in the relative share of rice, wheat and maize , while there is a decline in the relative share of jowar, bajra, ragi, barley and small millets. Thus, there is a shift towards the production of superior grains from coarse grains. There is also a shift towards greater production of cereals compared to pulses.

The output mix of cereal crops has undergone important changes over the years because of differential rates of growth recorded by component crops. Rice still remains the predominant cereal, though its position has been very much eroded by the emergence of wheat as a major crop. Similarly, maize has recorded gains while jowar has recorded substantial losses. Thus, changes have taken place in the pattern of production of foodgrains in India during the period under study. This is considered to be one of the important characteristic features of agricultural production in India.

(*g*) Index number of wholesale prices of total foodgrains in India increased from 35.20 in 1955-56 (with 1970-71 = 100) to 298.60 in 1986-87, *i.e.*, at a rate of 6.79 per cent per annum. Prices have shown a continued tendency to rise. Although foodgrains production has increased, prices of foodgrains have not fallen steeply because of the price support extended by Government through its procurement policy.

(*h*) Per capita net availability of foodgrains in India showed only a marginal increase from 430.9 grams per day in 1955-56 to 472.1 grams per day in 1986-87, *i.e.*, at a rate of 0.24 per cent per annum, with a peak level of 480.2 grams per day in 1964-65. The level of per capita net availability of foodgrains per day is lower than the minimum nutrition requirement of 594 grams per day, as specified by Indian Council of Medical Research.

(*i*) Procurement of foodgrains in India increased from 0.04 million tonnes in 1955-56 to 15.72 million tonnes in 1986-87, *i.e.*, at a rate of 14.09 per cent per annum. The foodgrains procured by the Government agencies are utilised for (*i*) running the Public Distribution System (PDS); (*ii*) issue to the roller flour mills for conversion into wheat products and (*iii*) building up of buffer stocks. The off-take of foodgrains from PDS increased from 2.08 million tonnes in 1955-56 to 18.37 million tonnes in 1986-87, *i.e.*, at a rate of 5.03 per cent per annum. The level of year-end closing stocks with the Government (Central and State Governments) increased from 0.32 million tonnes in 1955-56 to 23.36 million tonnes in 1986-87, *i.e.*, at a rate of 11.58 per cent per annum. The increase is found to be substantial from 1975 onwards.

(*j*) The biggest problem faced by India since Independence is that of growing food shortages. The Planning Commission accorded an important role to agriculture in the First Five-Year Plan, as the country was faced with serious food crisis and acute shortage of essential raw materials. Targets have been set for foodgrains production under successive Five-Year Plans. The analysis of targets and actual production of foodgrains in Seven Five-Year Plans of India indicates that there are divergencies between the targets and achievement of production of foodgrains in the successive Five-Year Plans of India, except in the First Five-Year Plan.

6.2.2. *Trends in the Imports of Foodgrains of India*

(*a*) Imports of foodgrains are represented by the imports of 'Cereal and Cereal Preparations', (Cereals for short). Imports of Cereals accounted for 4.53 per cent of the total imports of India in 1955-56. Cereals became the largest single item of imports in 1975-76 and accounted for more than a quarter of the total value of imports of India, exceeding the relative share of even the value of imports of POL. The relative share of imports of Cereals declined to 0.23 per cent in 1986-87 and to 0.15 per cent in 1987-88. Imports of Cereals come under the Group of 'Food and Live Animals Chiefly for Food'. Imports of Cereals accounted for 12.89 per cent of the imports of 'Food and Live Animals Chiefly for Food', in 1985-86. This brings out the relative importance of imports of Cereals in the total imports of India in general, and in the Group of imports of 'food and Live Animals Chiefly for Food', in particular.

(*b*) Imports of Cereals increased from Rs. 35.09 crores in 1955-56 to Rs. 523.10 crores in 1966-67, *i.e.*, at a rate of 17.69 per cent per annum during the pre-Green Revolution period. On the other hand, imports of cereals of India declined from Rs. 532.20 crores in 1967-68 to Rs. 47.00 crores in 1986-87, *i.e.*, at a rate of 4.12 per cent per annum, though not significant, during the post-Green Revolution period. On the whole, imports of Cereals in terms of value, increased from Rs. 35.09 crores in 1955-56 to Rs. 47.00 crores in 1986-87, *i.e.*, at a rate of 2.02 per cent per annum, but not significant. During this period, imports of Food and Live Animals Chiefly for Food increased from Rs. 57.76 crores in 1955-56 to Rs. 853.70 crores in 1985-86, *i.e.*, at a rate of 6.20 per cent per annum, and total imports of India increased from Rs. 1,972 crores to Rs. 20,201 crores *i.e.*, at a rate of 15.36 per cent per annum.

(*c*) Imports of foodgrains in terms of quantity increased substantially from 0.71 million tonnes in 1955-56 to 10.36 million tonnes in 1966-67, *i.e.*, at a rate of 17.43 per cent per annum, during the pre-Green Revolution period. Imports of foodgrains, in terms of quantity, declined from 8.67 million tonnes in 1967-68 to 0.13 million tonnes in 1986-87, *i.e.*, at a rate of 14.30 per cent per annum, during the post-Green Revolution period. On the whole, foodgrain imports of India in terms of quantity showed a decline from 0.71 million tonnes in 1955-56 to 0.13 million tonnes in 1986-87, *i.e.*, at a rate of 6.30 per cent per annum.

(*d*) Rice accounted for 37.27 per cent in terms of quantity and 13.39 per cent in terms of value of the total foodgrains imported by Indian in 1955-56. The relative share of rice declined to 2.33 per cent in terms of quantity and 2.13 per cent in terms of value in 1986-87. On the other hand, the relative share of imports of wheat increased in 61.18 per cent in 1955-56 to 83.72 per cent in 1986-87 in terms of quantity, and from 34.43 per cent in 1955-56 to 78.72 per cent in 1986-87 in terms of value. The relative share of other foodgrain imports increased from 1.55 per cent in 1955-56 to 13.95 per cent in 1986-87 in terms of quantity, but showed a decline from 52.18 per cent in 1955-56 to 19.15 per cent in 1986-87 in terms of value. Thus, the relative shares of rice, wheat and other foodgrains showed changes leading to changes in the pattern of foodgrain imports of India.

(*e*) Rice imports of India declined from 265 thousand tonnes in 1955-56 to 3 thousand tonnes in 1986-87, *i.e.*, at a rate of

11.09 per cent per annum. Imports of wheat declined substantially from 0.435 million tonnes in 1955-56 to 0.18 million tonnes in 1986-87, *i.e.*, at a rate of 6.53 per cent per annum.

(*f*) Apart from wheat and rice, India imported other foodgrains such as milo or sorghum, maize, other cereals unmilled, grams and other pulses from abroad. It is interesting to note that there is a change in the pattern of imports of other foodgrains from 1975-76 onwards. It indicates that instead of an increase in rice and wheat imports, there is an increase in pulses and other foodgrains imports.

(*g*) The analysis of direction of foodgrains imports indicates that the rice imports of India are met mainly by Burma, Thailand, UAR and USA, during the period 1955-56 to 1986-87. USA, Australia and Canada are the major suppliers of wheat imports of India.

(*h*) Several measures and policies have been adopted by the Government of India from time to time to reduce the imports of foodgrains and increase the domestic production of foodgrains. The import-availability ration (M/S) of foodgrains indicates the progress of import substitution that has taken place over time as a result of the measures taken by the Government. The average M/S ratio for total foodgrains declined from 0.03 during 1955-60 to 0.0001 during 1985-87. The above analysis indicates that imports of foodgrains accounted for a relatively small proportion of the total supplies. There is a reduction in the degree of import dependence of India on foodgrain imports during the period under study, *i.e.*, 1955-56 to 1986-87.

Imports of foodgrains were mostly undertaken to meet the shortfalls in the production of foodgrains in relation to the growing needs of the economy. In recent times, foodgrain imports are also being undertaken to build up buffer stocks.

6.2.3. *Determinants of Imports of Foodgrains of India*

(*a*) The major factors considered in determining the demand for imports of foodgrains of India are: per capita income, relative price, *i.e.*, the ratio of import price to domestic price, domestic production of foodgrains, stocks of foodgrains lagged by one year, procurement of foodgrains, external assistance and net foreign assets of India. Imports of foodgrains of India are assumed to be inversely related to relative price, domestic production of foodgrains, procurement of foodgrains and directly related to per capita income, external assistance and net foreign assets. To capture the effects of policy changes of the Government, dummy variables are also included in the import demand function of foodgrains of India.

(*b*) The analysis of the import demand function for foodgrains indicates that the variables, *viz.*, per capita income, relative price, production of foodgrains, stocks of foodgrains lagged by one year, procurement of foodgrains, external assistance, net foreign assets and dummy variables together explained 73.73 per cent of the variation in the demand for imports of foodgrains of India during the period 1955-56 to 1986-87. Among the variables considered, the coefficients of the variables *viz.*, per capita income, production of foodgrains and stocks of foodgrains lagged by one year are found to be statistically significant, with expected signs. From the analysis of the empirical results, it is inferred that a one unit increase in the per capita income of India leads to 0.06 units increase in the demand for imports of foodgrains of India. A one unit increase in the domestic production of foodgrains, on the other hand, leads to 0.16 units decline in the imports of foodgrain of India. Imports of foodgrains are also found to be influenced by buffer stock policy and procurement policy of the Government. The impact of the policies of the Government are found to be more in the first period 1955-56 to 1974-75 than the second period 1975-76 to 1986-87.

(*c*) From this it may be concluded that per capita income, domestic production of foodgrains, stocks of foodgrains in the previous period and the policies of the Government regarding maintenance of buffer stocks and procurement of foodgrains are the important factors determining the demand for imports of foodgrains of India, during the period 1955-56 to 1986-87.

(*d*) It may be mentioned here that external assistance and net foreign assets are not found to be significant determinants of the demand for foodgrain imports of India during the period under study. This may be due to the fact that food imports, being essential imports, foreign exchange constraint may not operate in the same way as for non-essential imports.

6.3. Policy Implications

Cereals like rice, wheat and millets constitute the basic food of human beings. Food, drinking water, elementary education and primary health services are identified by the World Bank as the basic minimum needs of the poor sections of population in developing countries like India.

Production of foodgrains in India increased from 69.3 million tonnes in 1955-56 to 143.4 million tonnes in 1986-87, *i.e.*, at a rate of 2.55 per cent per annum. However, the per capita net availability of foodgrains in India showed only a marginal increase from 430.9 grams per day in 1955-56 to 472.1 grams per day in 1986-87, *i.e.*, at a rate of 0.24 per cent per annum, with a peak level of 480.2 grams per day in 1964-65. The level of per capita net availability of foodgrains per day is lower than the minimum nutrition requirements of 594 grams per day, as specified by the Indian Council of Medical Research (ICMR).

It is, therefore, necessary for the Government to formulate policies, which would help to increase the per capita net availability of foodgrains in India in order to reach the minimum nutrition requirements as prescribed by ICMR. Some of the measures that could be taken by the Government are indicated below.

(a) Imports of foodgrains have to be continued till such time that the per capita net availability of foodgrains is improved substantially in order to reach the minimum nutrition requirements as prescribed by the ICMR. This has to be supplemented by measures to strengthen the public Distribution System (PDS) of foodgrains in India.

(*b*) The present study indicates that the major factors determining imports of foodgrains of India are: domestic production of foodgrains, stocks of foodgrains, procurement of foodgrains and the Government policies relating to the management of buffer stocks and procurement of foodgrains. From this, it may be inferred that a crop failure, a decline in the buffer stocks of foodgrains or insufficient procurement of foodgrains further enhances the need for imports of foodgrains to maintain the minimum nutritional requirements of the vulnerable sections of the society.

(*c*) Increase in the production of foodgrains in India has contributed to the achievement of food security but has not increased the economic access to food. Therefore, steps have to be taken by the Government to imporve the income distribution as well as to pursue to rational food price policy to minimise the problems of mal-nourishment, hunger and under-nourishment that continue to plague the economy.

(*d*) Production of foodgrains in India has shown considerable fluctuations, which are found to be mostly arising from instability in the yield per hectare of foodgrains. The greatest single limiting factor is uncertain availability of water and the dependency on erratic monsoons. Hence, Government has to take steps to imporve the irrigation facilities in the economy in the long run as well as to reduce to gap in the utilisation of the created irrigation potential.

It is proposed to increase the irrigation potential from 67.5 million hectares at the end of the Sixth Five-Year Plan to 80.4 million hectares by the end of the Seventh Five Year Plan. In this context, it is highly essential to ensure the efficient use of the available irrigation facilities in order to stabilise as well as to increase the production of foodgrains and thereby to improve the per capita availability of foodgrains in India in the long run.

APPENDIX

Appendix Table 5.1 : Zero Order Correlation Matrix of Import Demand Function of Foodgrains of India 1955-56 to 1986-87

	Q_{mft}	PY_t	RP_{ft}	P_{ft}	$S_{f(t-1)}$	PR_{ft}	EA_t	NFA_t	D	$DS_{f(t-1)}$	DP
Q_{mft}	1.00	-0.46	-0.52	-0.46	-0.63	-0.38	-0.41	-0.58	-0.42	-0.58	-0.46
PY_t	-0.46	1.00	0.47	0.98	0.89	0.95	0.89	0.92	0.85	0.87	0.91
RP_{ft}	-0.52	0.47	1.00	0.49	0.66	0.43	0.55	0.70	0.68	0.72	0.64
P_{ft}	-0.46	0.98	0.49	1.00	0.84	0.97	0.84	0.89	0.86	0.83	0.91
$S_{f(t-1)}$	-0.63	0.89	0.66	0.84	1.00	0.80	0.85	0.95	0.83	0.97	0.83
PR_{ft}	-0.38	0.95	0.43	0.97	0.80	1.00	0.84	0.87	0.85	0.79	0.91
EA_t	-0.41	0.89	0.55	0.84	0.85	0.84	1.00	0.89	0.79	0.87	0.86
NFA_t	-0.58	0.92	0.70	0.89	0.95	0.87	0.89	1.00	0.91	0.97	0.94
D	-0.42	0.85	0.68	0.86	0.83	0.85	0.79	0.91	1.00	0.88	0.96
$DS_{f(t-1)}$	-0.58	0.87	0.72	0.83	0.97	0.79	0.87	0.97	0.88	1.00	0.88
DP	-0.46	0.91	0.64	0.91	0.83	0.91	0.86	0.94	0.96	0 .88	1.00

SELECT BIBLIOGRAPHY

Agarwala, R. *An Econometric Model of India: 1948-1961*, Frank Cass & Co, Ltd., London, 1970.

Aggarwal, M. "Devaluation, Determinants Of International Trade Flows And Payments Imbalance", *The Indian Economic Journal*, Vol. 31, No.3, Jan-March, 1984

Agarwal, A.N. *Indian Economy*, Allied Publishers, New Delhi, 1975

Alderman, H.and Braun, J.V. *The effects of the Egyptian Food Ration And Subsidy System On Income Distribution And Consumption*, International Food Policy Research Institute (IFPRI) Research Report-45, July, 1984.

Alejandro, D. *Exchange Rate Devaluation In A semi-Industrial Country: The Experience of Argentina 1955-1961*, Cambridge, Massachusetts, 1965.

Balassa, B. *Trade Prospects For Developing Countries*, Yale University Press, New York, 1964.

Bansil, P.C. *Agricultural Problems of India*, Oxford and IBH Publishing Co., New Delhi, 1981.

Behari, B. *Imports In A Developing Economy*, Vora & Co., Publishers, Bombay, 1965.

Benarji, H. "Analysis of Import Demand", *Artha Vijnana*, Vol. I, No. 3, September, 1959.

Bhagwati, J.N. and Desai, P. *India: Plannning For Industrialisation*, Oxford University Press, London, 1970.

Bhalla, G.S. and Alagh, Y.K. *Performance of Indian Agriculture: A District-wise study*, Sterling Publishers Pvt. Ltd., New Delhi, 1979

Bhatia, B.M. *India's Food Problem And Policy Since Independence*, Somaiya Publications Pvt. Ltd., Bombay, 1970.

Biswas, B. and Ram, R. "Demand Function for India's Foodgrain Imports: Some Elasticity Estimates", *The Indian Economic Journal*, Vol. 27, No. 4, April-June, 1980.

Byerlee, D. "The Political Economy of the Third World Food Imports: The Case of Wheat", *Economic Development and Cultural Change*, Vol. 35, No.2, January 1987.

Chandra, S.C. *Agricultural Price Policy In India*, Chugh Publications, Allahabad, 1985.

Chatterji, S.K. *Population and Food: Indian Economy Since Independence Series*, National Publishers, Calcutta, 1969.

Dantwala, M.L. "Agrarian Structure and Economic Development", *Indian Journal of Agricultural Economics*, Vol. XV, No.1, January-March 1961.

De Janvry, A. and Subba Rao, K. *Agricultural Price Policy and Income Distribution in India*, Oxford University Press, Delhi, 1986.

Diwan, R.K. "Import Structure and Policy", *Foreign Trade Review*, Vol. No.1, April-June 1968.

Dutta, M. "A Prototype Model of India's Foreign Sector", *International Economic Review*, 1964.

Dutta, M. "Import Structure of India", *Review of Economics and Statistics*, Vol. XL-VII, 1965.

Dutta, M. "Measuring Role of Price in International Trade: Some Further Tests", *Econometrica*, Vol. XL-VII, July 1965.

Dutt, R. and Sundram, K.M. *Indian Economy*, S. Chand and Company Ltd., New Delhi, 1986.

E.S.R.F. *Structure of Indian Imports 1957-64*, Economic and Scientific Research Foundation, Paper-4, New Delhi. 1966.

Etienne, G. *Food And Poverty: India's Half Won Battle*, Sage Publications, New Delhi, 1988.

Gavan, J.D. and Chandrasekhara, I.S. *The Impact of Public Foodgrain Distribution on Food Consumption And Welfare in Srilanka*, International Food Policy Research Institute (IFPRI) Research Report 13, December 1979.

George, P.S. *Public Distribution of Foodgrains In Kerala-Income Distribution Implications And Effectiveness*, International Food Policy Research Institute (IFPRI) Research Report-7, March 1979.

Govt. of India (G.O.I) *All-India Agricultural Labour Inquiry 1950-51*. 1951.

Bulletin On Food Statistics, Various issues, Delhi.

Govt. of India (C.S.O) Central Statistical Organisation, *Statistical Abstract: India*, various issues, Delhi.

Govt. of India (DGCI & S) Directorate General of Commercial, Intelligence and Statistics, *Monthly Statistics of Foreign Trade of India*, various issues, Calcutta.

Govt. of India (G.O.I) *Economic Survey*, various issues, Delhi.

——— *Economic Survey of Indian Agriculture 1966-67*, Delhi.

——— *Estimates of Area And Production of Principal Crops of India*, various issues, Delhi.

——— *Food Corporation of India (FCI)*, various Reports, Delhi.

——— *The Foodgrains Policy Committee*, Delhi, 1969.

——— Ministry of Commerce and Industry, Annual *Administration of The Import And Export Trade Control*, 1962, Delhi.

——— Planning Commission, *Seventh Five-Year Plan (1985-86)*, Vol. I, Nov. 1985, Delhi.

——— Planning Commission, *Plan Documents*, various issues, New Delhi.

Gulati, A. *Agricultural Price Policy in India: An Econometric Approach*, Concept Publishing Company, New Delhi, 1987

Haberler, G. "Terms of Trade and Economic Development", In Ellis, H. (ed.), *Economic Development of Latin America*, St. Martin's Press, New York, 1961.

Hemphill, W.L. "The Effect of Foreign Exchange Receipts on Imports of Less Developed Countries", *IMF Staff Papers*, Nov. 1974.

Hicks, J.R. *Capital And Growth*, Oxford University Press, London, 1965.

Hirschman, A.O. *The Strategy of Economic Development*, Yale University Press,.. New Haven, 1958.

I.C.M.R. Indian Council of Medical Research, *Recommended Balanced Diet*, 1966.

Islam, B "Price, Income and Foreign Exchange Reserve Elasticity for Asian Rice Imports", *American Journal of Agricultural Economics*, Vol. 60, No.3, August, 1978.

Islam, N. "Experiments in Econometric Analysis of an Import Demand Function", *Pakistan Economic Journal*, September 1961.

Jayaraman, T.K. "Import Demand Function of India", *Foreign Trade Review*, Vol. XI, No.4, January-March, 1977.

Jain, S.C. *Agricultural Price Policy in India*, Himalaya Publishing House, Bombay, 1986

Johnston, B.F. and Mellor, J.W. "The Role of Agriculture In Economic Development", *American Economic Review*, Vol. LI, No.4. September, 1961.

Joseph, S.C. *Food Policy and Economic Development in India*, Allied Publishers Pvt. Ltd., Bombay, 1961.

Khan, M.S. "Import and Export Demand in Developing Countries", *IMF Staff Papers*, November 1974.

Kindleberger, C.P. *Foreign Traded and the National Economy*, Yale University Press, New Haven, 1962.

Kindleberger, C.P. and Herrick Bruce. *Economic Development*, McGraw Hill, Kogakusha Ltd., New York, 1977.

Kulshreshtha, S.N. and Tiwari, D.D. "Agriculture in India-Problems and Prospects", in Uppal, J.S. (ed.), *India's Economic Problems:An Analytical Approach*, Tata McGraw-Hill Publishing Company Ltd., New Delhi, 1981.

Kuznets, S. *Six Lectures on Economic Growth*, Free Press, Glencoe, 1959.

Kuznets, S. "Economic Growth and the Contribution of Agriculture: Notes on Measurement", *International Journal of Agrarian Affairs*, Vol.3, No.2, April 1961.

Lekhi, R.K. *Agricultural Development in India,* Classical Publications Company, New Delhi, 1986.

Lewis, W.A. *The Theory of Economic Growth,* Allen and Unwin, London, 1955.

Madalgi, S.S. *Population and Food Supply in India*, Lalvani Publishing House, Bombay 1970.

Maisels, A. *Exports And Economic Growth of Developing Countries,* Cambridge, 1968.

Marwah, K. "An Econometric Model of India: Estimating Prices, Their Role and Sources of Change", *Indian Economic Review,* Vol. VII, No.1, April, 1972.

Mehra, S. *Instability in Indian Agriculture in the Context of the New Technology,* International Food Policy Research Institute (IFPRI), Report 25, July 1981.

Merier, G.M. *International Economics: The Theory and Policy,* Oxford University Press, New York, 1980.

Mellor, J.W. *The Economics of Agricultural Development*, Vora & Co., Bombay, 1969.

Mellor, J.W. "Food and Development: The Critical Nexus", *Economic Impact,* No.61, May 1987.

Mitra, A.K. Food Problem in India: A *Quantitative Study in the Regional Perspective,* Anu Books, Meerut, 1986.

Morton, K. and Tulloch, P. *Trade and Developing Countries,* Croom Helm, London, 1977.

Moorthy, V.N. and Sastry, V.K. "Elasticities of Demand for Certain Indian Imports and Exports", *Sankhhya* Vol. VIII, December 1951.

Narshimham, N.V.A. *A Short Term Planning Model for India*, North Holland Publishing Company, Amsterdam, 1956.

Nurkse, R. "Patterns of Trade and Development", Wicksell Lectures (1959) in Harberler, G. and Stern, R.M. (ed.), *Equilibrium and Growth in the World Economic,* Harvard University Press, Cambridge, Massachusetts, 1961.

Panchamukhi, V.R. "A Survey of Econometric Studies in the Fields of Production, International Trade, Money and Banking", *A Survey of Research in Economics*, ICSSR, Allied Publishers, Delhi, 1978.

Panchamukhi, V.R. *Trade Policies of India: A Quantitative Analysis*, Concept Publication Co., Delhi, 1978.

Panse, V.G. "Trends in Area and Yield of Principal Crops in India",*Agricultural Situation of India*, June 1952.

Parthasarathy, G. and Mudhahar, M.S. "Foodgrain Prices and Economic Growth", *Indian Journal of Agricultural Economics*, Vol. 31, No.2, April-June 1976

Patel, S.K. *Quarterly Econometric Analysis of Imports in India*, Hindustan Publishing Corporation, Delhi, 1981.

Patel, S.K. "Relative Prices in Indian Import Functions", The Indian Economic Journal, Vo. 23, April-June, 1976.

Paul, S. What Price Food?. *Agricultural Price Policies in Developing Countries*, St. Martin's Press, New York, 1987.

Ramachandran, L. *India's Food Problem*: A New Approach, Allied Publishers Pvt. Ltd., New Delhi, 1977.

Ramesh Babu, S. *India's Foreign Trade: Some Trends*, Chugh Publications, Allahabad, 1988.

Rao, M.S. *Indian Economy*, Premier Book Company, Delhi, 1970.

Rao, V.K.R.V. Food, Nutrition And Poverty in India, Vikas Publishing House Pvt. Ltd., Delhi 1982.

Rao, V.M. *Second India Studies: Food*, The MacMillian Company of India Ltd., Delhi, 1975.

Reserve Bank of India (RBI) *Report on Currency and Finance*, various issues, Bombay.

Reutlinger, S. "The Level of Stability of India's Foodgrain Consumption" in World Bank, *India: Occasional Papers*, No.279, November 1978.

Ricardo, D. *Principles of Political Economy And Taxation*, the Mac Millian Press Ltd., London, 1817.

Ricardo, D. *Principles of Political Economy*, Sraffa Edition, London.

Rostow, W.W. *Stages of Economic Growth*, Cambridge University Press, London, 1962.

Samuleson, P.A. *Economics*, McGraw Hill, Kogakusha Ltd., 1973.

Sarma, J.S. "India-A Drive Towards Self-sufficiency In Foodgrains", *American Journal of Agricultural Economics*, Vol. 60 No.5, December 1978.

Sarma, J.S. Shyamal, R. and George, P.S. *Two Analyses of Indian Foodgrain Production and Consumption Data*, International Food Policy Research Institute (IFPRI) Research Report, 12, November 1979.

Sawant, S.D. "A Review of Performance in the Agricultural Sector", in Brahmananda, P.R. and Panchamukhi, V.R. (ed.), *The Development Process of the Indian Economy*, Himalaya Publishing House, Bombay, 1987.

Sharma, O.P."Foreign Trade and Relative Prices in an Import Substituting Economy", *Indian Economic Journal*. January-March, 1975.

Schultz, T.W. *The Economic Organisation of Agriculture*, McGraw Hill, New York, 1953.

Schultz, T.W. "A New Era For Agriculture in Economic Growth", *Indian Journal of Agricultural Economics*, Vol. XIV, No.4, October-December, 1959.

Smith, A. *Wealth of Nations*, Modern Library Edition, New York, 1913.

Singh, M. "Trade Policies and Economic Development" in *Economic Development in South Asia*, (ed. by Robinson, E.A.G. and Kidron, M., Proceedings of a Conference held by The International Economic Association at Kandy, Ceylon, Mac Millian, St. Martin's Press, London, 1970.

Sinha, R.P. *Food In India: An Analysis of the Prospects for Self Sufficiency by 1975-76*, Oxford University Press, Bombay, 1961.

Stamp, E. *Growing Out of Poverty*, Oxford University Press, Oxford, 1977.

Sukhatme, P.V. *Feeding India's Growing Millions*, Asia Publishing House, 1965.

U.N. *Yearbook of International Trade Statistics*, various issues, New York.

U.N., Escap *Committee on Agricultural Development Special Study on Food Supply and Distribution System*, Bangkok, January, 1982.

U.N.,F.A.O. *6 Billion To Feed*, Rome, 1962.

U.N., F.A.O. Commodity Bulletin Series, *The Economic Relationship Between Grain and Rice*, Report 39, U.S.A.1965.

U.N.,F.A.O. *Trade Yearbook*, various issues, Rome.

U.N. World Bank, *India: Occasional Papers*, World Bank Staff Working Paper No.279, November 1978.

Upadhyaya, K.K. *Import Financing in India*, Chugh Publications, Delhi, 1980.

Verma, P.C. "Import Structure of India, Bangladesh and Pakistan", *Foreign Trade Review*, Vol. XII, No.4, January-March 1978

Viner, J. *International Trade and Economic Development*, Calrendon Press, Oxford, 1952.

Wadia, F.K. "Agricultural Price Policy", in J.N. Mongia (ed.), *India's Economic Policies 1947-77*, Allied Publishers Pvt. Ltd., Delhi, 1980.

Wadhwa, C.D. "Estimates of Demand for Indian Exports and Imports and the Question of the Devaluation of the Indian Rupee", *Foreign Trade Review*, Vol. VIII, No.4, January-March 1974.

Wall, J. "Foodgrain Management: Pricing, Procurement, Distribution, Imports and Storage Policy", World Bank, *India: occasional Papers*, Staff Working Paper No.279. November, 1978.

Yudelman, M. *Agricultural Development and Economic Integration in Latin America*, George Allen and Unwin Ltd., London, 1970.

INDEX